Engineering Drawing Fundamentals

M. A. Parker, MEd, IEng, MIMechIE
Senior Technical Education Officer, British Standards Institution

L. J. Dennis, BSc(Eng), DipNEC
Technical Adviser to the Joint Matriculation Board

STANLEY THORNES (PUBLISHERS) LTD.

First Published 1992 by:
Stanley Thornes (Publishers) Ltd.
Old Station Drive
Leckhampton
CHELTENHAM GL53 0DN
England

A catalogue record for this book is available from the British Library.

ISBN 0 7487 1559 2

Typeset by Tech-Set, Gateshead, Tyne & Wear
Printed and bound in Great Britain at The Bath Press, Avon.

Contents

Preface

This book has been written to cover the engineering drawing material in the BTEC unit Engineering Fundamentals. The aim of the authors has been to provide a book which presents the basic principles of the engineer's language without unnecessary elaboration. The text has been supplemented with numerous examples, some of which are laid out in step-by-step form.

Numerous exercises have been provided, but it is not the authors' intention that all of these should be worked by every student in the class. Rather, it is hoped that there are sufficient to give variety in assignments for a class and to enable the lecturer to set phase tests as considered necessary.

As well as the exercises, many self-assessment questions are included. Answers to these are given together with solutions to a selection of the exercises. In addition, a post test follows each chapter.

The book draws material extensively from BS 308: *Engineering drawing practice*. In the chapter on electrical circuit diagrams the symbols are taken from BS 3939: *Graphical symbols for electrical power, telecommunications and electronics diagrams*. The British Standards Institution kindly gave permission for extracts from these standards to be reprinted.

M. A. Parker
L. J. Dennis
London 1992

Orthographic projection

Principal learning objectives

The student explains with the aid of drawings the meaning of orthographic projection in terms of the principal planes of projection, points, lines, areas and simple three-dimensional objects. The student produces neat and clear drawings, including the use of hidden detail lines, using first- and third-angle projection.

Introduction

The language of engineering is engineering drawing. It conveys information about engineering parts to everyone who will be involved in their manufacture. The information should be all that is necessary to make the parts, and should be shown with clarity and precision and in a way which is universally understood. These objectives may be met by making sure that drawings follow the recommendations of British Standard 308, *Engineering drawing practice*. An abridgement of this standard, PP 7308, *Engineering drawing practice for schools and colleges*, has been produced by the British Standards Institution, and the information in it is sufficient for most engineering drawing courses.

All drawings, whatever their type, have to solve satisfactorily the basic problem of representing a three-dimensional object on a two-dimensional sheet of paper. In engineering, the problem is solved generally by using **orthographic projection**, the fundamentals of which are explained in this chapter. **Pictorial projections** are sometimes used. These are explained in Chapter 7. Orthographic projection is preferred because it is easier to draw than a pictorial projection, and gives accurate information about the shape and size of an object in a way that is impossible with words.

Orthogonal projectors

From points

Figure 1.1 shows three points A, B and C in space, each point being adjacent to a plane. Lines are

shown from the points to the planes, meeting them at a, b and c respectively. The lines are **projectors**, and a, b and c are projections of A, B and C on the planes. If the projectors are normal to the planes, that is, perpendicular to them, then they are **orthogonal**, which means 'drawn at right angles'. Similarly, a, b and c are **orthogonal projections** of A, B and C.

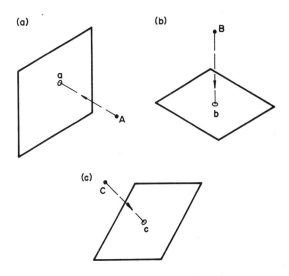

Aa, Bb and Cc are projectors normal to planes

Figure 1.1 *Orthogonal projection of a point*

From lines

A line may be projected orthogonally by projecting its ends, as shown in Figure 1.2. The projectors will be parallel and at right angles to the plane. Note that the projection will show the **true length** of the line only if the line is parallel to the plane, as with line CD.

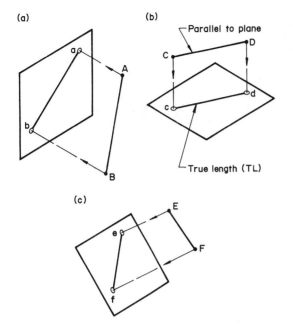

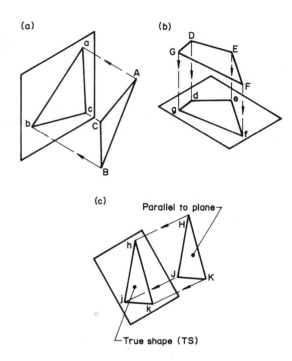

ab, cd and ef are orthogonal projections of lines AB, CD and EF. The projectors are parallel and perpendicular to the planes of projection

Figure 1.2 *Orthogonal projection of a line*

abc, defg and hjk are orthogonal projections of areas ABC, DEFG and HJK. The projectors are parallel and perpendicular to the planes of projection

Figure 1.3 *Orthogonal projection of an area*

From areas

An area may be projected orthogonally by projecting the lines forming its boundaries, as illustrated in Figure 1.3. Again, the projectors will be parallel and at right angles to the plane of projection. If the area is parallel to the plane, as is the case with triangle HJK, then the projection will be the **true shape** of the area. It follows that all the sides of the triangle will then be true lengths.

From solids

The sides of a solid are areas, so using sets of parallel projectors the solid may be projected onto any number of planes. This is shown in Figure 1.4. As before, the projectors in each set are parallel and at right angles to the projection plane.

In Figure 1.4(a) and (b), one face of the solid is parallel to the plane of projection. Therefore, that face will be shown as its true shape and its boundaries will be true lengths.

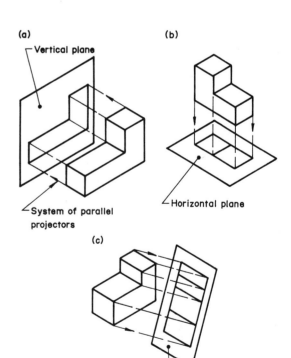

Figure 1.4 *Orthogonal projection of a solid*

Self-assessment questions

1.1 What advantages does orthographic projection have over pictorial projections?

1.2 What is an orthogonal projector?

1.3 How must a square be positioned relative to a plane for its projection on the plane to be a square?

Orthographic projection

If views of an object are projected orthogonally onto a vertical and a horizontal plane, then those views, taken together, are said to be in **orthographic projection**. The vertical and horizontal planes are the **principal planes of projection**. They intersect producing four quadrants or **angles**, as shown in Figure 1.5, but only two, the first and third, are used in practice.

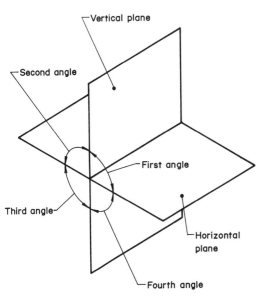

Figure 1.5 *Principal planes of projection*

The object to be drawn is imagined to be placed in one of these quadrants, and views of it are projected orthogonally onto the planes. These are the **principal views** of the object. The object may be placed in any position relative to the planes, but normally its main faces are parallel to them. As explained earlier, this ensures that views of the faces are true shapes and their boundaries are true lengths. This means that dimensions applied to the views will be true lengths, thus giving an accurate picture of the size and shape of the object.

First-angle orthographic projection

Figure 1.6 shows an object positioned in space in the first-angle quadrant. Views of the object have been drawn on the planes using sets of parallel projectors perpendicular to the planes. The view on the vertical plane is called the **elevation**, that on the horizontal plane is called the **plan**. To obtain the views as they would appear on a sheet of paper, the horizontal plane is opened out about the intersection line between the planes. This line is called the **XY line**, **ground line** or **folding line**. Relative to the elevation it represents the horizontal plane; relative to the plan it represents the vertical plane.

The lower half of Figure 1.6 shows the plan and elevation of the object as they would appear on a sheet of paper. It shows that in first-angle orthographic projection the elevation appears above the plan. On a drawing prepared for use in industry, only the elevation and plan would be shown, the planes, XY line and projectors being omitted.

Third-angle orthographic projection

An object positioned in the third-angle quadrant is shown in Figure 1.7. Since the planes now come between the observer and the object they are considered to be transparent, and the object is viewed through them. An elevation and plan have been projected orthogonally onto the planes. When the horizontal plane is opened out about the ground line the views appear as in the lower half of Figure 1.7, with the plan now above the elevation.

Projection symbols

BS 308 states that both first-angle and third-angle projections 'are approved internationally and have equal status'. Therefore it is necessary to indicate on a drawing which system has been used. This is done by a symbol consisting of two views of a frustum of a

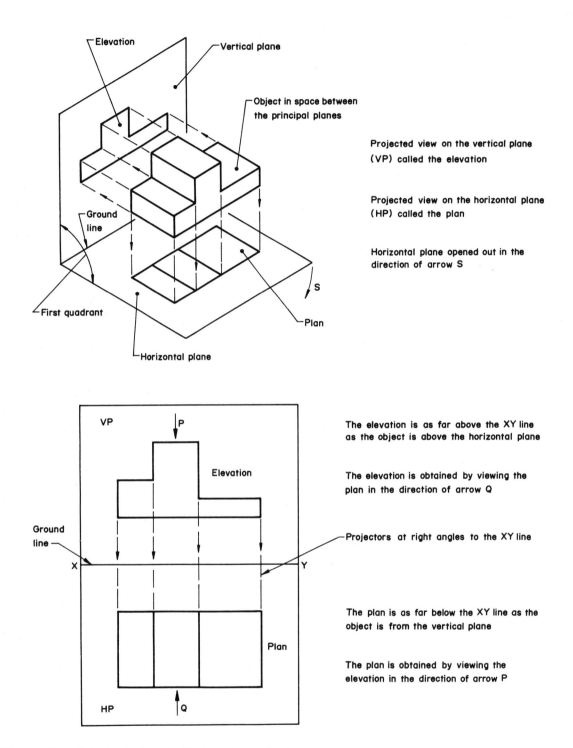

Projected view on the vertical plane (VP) called the elevation

Projected view on the horizontal plane (HP) called the plan

Horizontal plane opened out in the direction of arrow S

The elevation is as far above the XY line as the object is above the horizontal plane

The elevation is obtained by viewing the plan in the direction of arrow Q

Projectors at right angles to the XY line

The plan is as far below the XY line as the object is from the vertical plane

The plan is obtained by viewing the elevation in the direction of arrow P

Figure 1.6   *Principal views in first-angle projection*

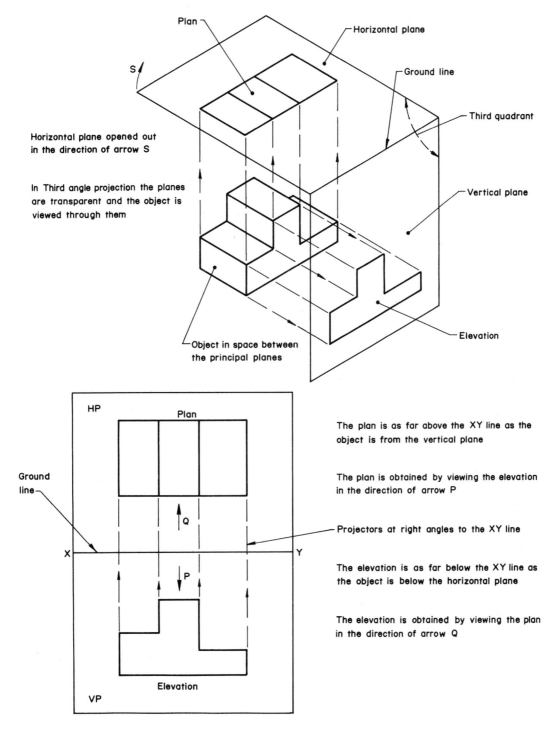

Plan

Horizontal plane

S

Ground line

Third quadrant

Horizontal plane opened out
in the direction of arrow S

In Third angle projection the planes
are transparent and the object is
viewed through them

Vertical plane

Object in space between
the principal planes

Elevation

HP

Plan

Ground
line

Q

X   Y

P

Elevation

VP

The plan is as far above the XY line as the
object is from the vertical plane

The plan is obtained by viewing the elevation
in the direction of arrow P

Projectors at right angles to the XY line

The elevation is as far below the XY line as
the object is below the horizontal plane

The elevation is obtained by viewing the plan
in the direction of arrow Q

Figure 1.7   *Principal views in third-angle projection*

cone. Figure 1.8(a) shows the recommended proportions for the symbol and Figure 1.8(b) and (c) show the symbols for first- and third-angle projection respectively.

Failure to identify the projection system may result in costly errors being made by a person, used to one system, misreading a drawing made in the other.

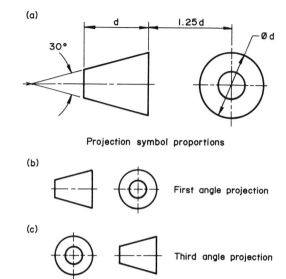

Projection symbol proportions

(b) First angle projection

(c) Third angle projection

Figure 1.8   *Projection symbols*

Self-assessment questions

1.4   Indicate whether the following statements are true or false. If requested, give a written statement or a sketch to justify your answers.

   (a)   A folding line is the line of intersection between two adjacent projection planes.

   (b)   Views of an object are different if they are drawn in first-angle and third-angle projection.

   (c)   A plan view is obtained if an object is viewed from above.

   (d)   In an elevation the XY line represents the horizontal plane.

Exercise 1.1

PROJECTION

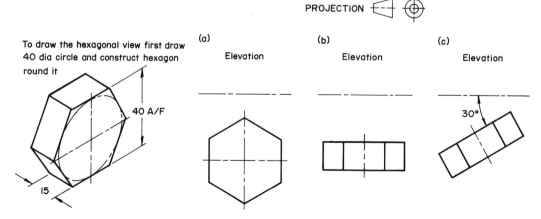

To draw the hexagonal view first draw 40 dia circle and construct hexagon round it

40 A/F

15

(a) Elevation

(b) Elevation

(c) Elevation

30°

Draw the given plan views of the hexagonal prism and project the elevations.

Exercise 1.2

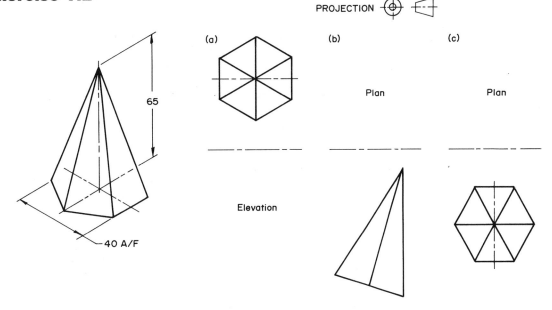

PROJECTION ⊕ ◁

65

40 A/F

(a)

(b)   Plan

(c)   Plan

Elevation

Draw the given views of the hexagonal pyramid and project the second view. In (b) a triangular face is vertical.

The projection of end views

Sometimes it is found that an elevation and a plan of an object are not sufficient to describe it completely, or they are ambiguous. For example, Figure 1.9(a) shows the elevation and plan of a block. From these views four possible interpretations of the shape of the block are possible, as shown by the pictorial views (b), (c), (d) and (e). To show which is the correct interpretation, a third view, called an **end view**, **end elevation** or **side elevation**, is necessary. This view is obtained by viewing the elevation in the direction of arrow X.

The end view is drawn on an **auxiliary vertical plane** which is at right angles to both the horizontal and vertical planes. This plane may be placed in either of the positions shown in Figure 1.10, in which the views are in first-angle projection. The

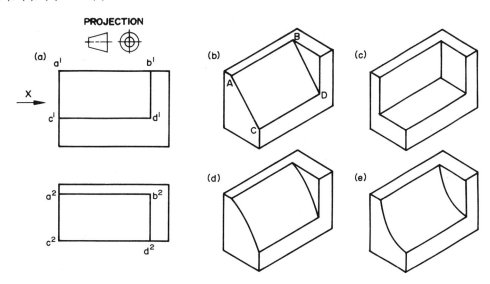

The elevation and plan are the same for all four objects

Figure 1.9   *Ambiguous views*

7

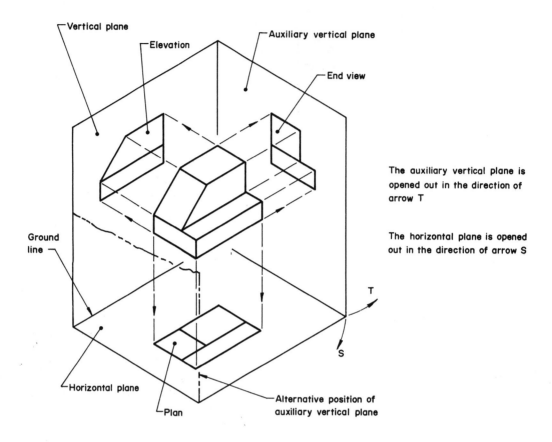

The auxiliary vertical plane is opened out in the direction of arrow T

The horizontal plane is opened out in the direction of arrow S

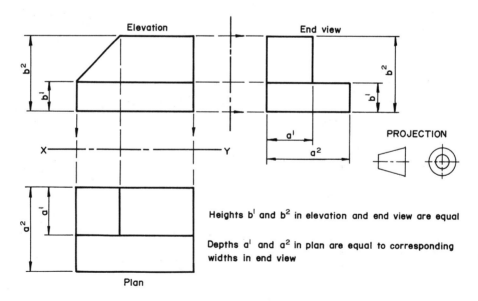

Heights b^1 and b^2 in elevation and end view are equal

Depths a^1 and a^2 in plan are equal to corresponding widths in end view

Figure 1.10   *Projection of end views in first-angle projection*

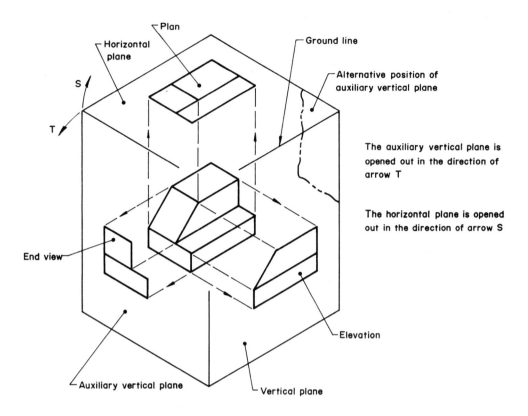

The auxiliary vertical plane is opened out in the direction of arrow T

The horizontal plane is opened out in the direction of arrow S

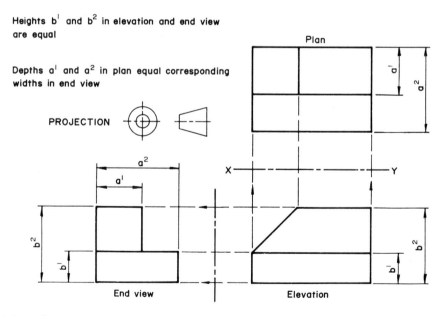

Heights b^1 and b^2 in elevation and end view are equal

Depths a^1 and a^2 in plan equal corresponding widths in end view

Figure 1.11 *Projection of end views in third-angle projection*

position chosen for the auxiliary vertical plane depends on which end face of the object is the more important. If neither face is more important than the other, the end view to be drawn is that which shows the minimum of hidden detail (see below). In some cases it may be necessary to draw both end views.

When the auxiliary vertical plane is opened out with the horizontal plane, the three views appear in the positions shown in the lower half of Figure 1.10. It can be seen that heights in the elevation and end view are equal, and depths in the plan view are equal to corresponding widths in the end view. This means that a third view can be projected from any two other views. Note that any orthographic view shows only two of the dimensions height, length and width of the object.

In Figure 1.11 the projection of an end view in third-angle projection is shown. The auxiliary vertical plane is now transparent and when it is opened out with the horizontal plane the views occupy the positions shown in the lower half of Figure 1.11. Now the plan and end views are placed at the sides of the front elevation nearest to the faces which they represent. In first-angle projection they are placed at the sides of the front elevation remote from the faces which they describe. Thus, for long objects, third-angle projection has the advantage that end views appear nearest to the faces which they represent.

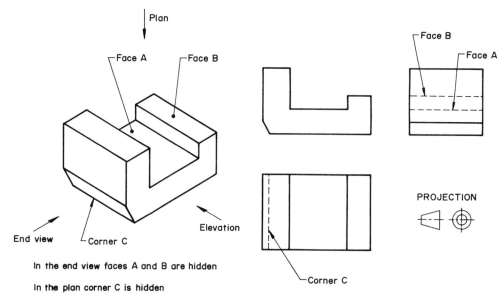

In the end view faces A and B are hidden

In the plan corner C is hidden

Figure 1.12 *Hidden faces and corners*

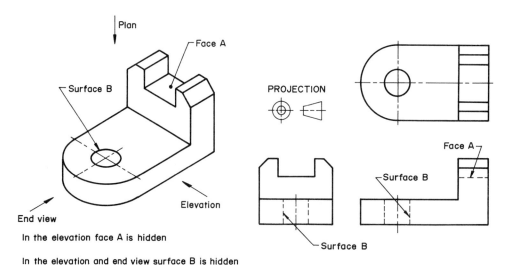

In the elevation face A is hidden

In the elevation and end view surface B is hidden

Figure 1.13 *Hidden faces and holes*

Visible lines, hidden lines and centre lines

(a) CONTINUOUS THICK ——————— Visible outlines and edges

(b) DASHED THIN — — — — — — — — Hidden outlines and edges. Dashes and gaps to be of consistent lengths

(c) CHAIN THIN ——— — — ——— Centre lines, lines of symmetry, pitch lines and pitch circles
    Dashes and gaps to be of consistent lengths. These lines begin and end with a long dash. Long dashes cross.

Thick lines are twice as wide as thin lines. Commonly used widths are 0.7 mm and 0.35 mm

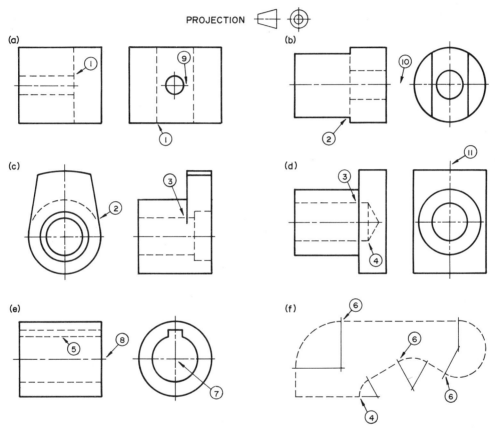

I   Hidden lines touch outlines and other hidden lines

2   Gap where hidden line continues outline

3   Where a hidden line crosses an outline a gap is left each side of the outline

4   Hidden lines meet at a point

5   Gaps staggered in parallel hidden lines

6   Hidden arcs stop at tangent points

7   Centre lines cross at long dashes

8   Long dash at ends of centre line extends a short distance past the outline

9   For small circles (up to about 12 mm diameter) centre lines are continuous thin lines

10 Centre lines do not extend across the space between views

11 Line of symmetry of the rectangular base of the part

Figure 1.14   *Treatment of hidden lines and centre lines*

In third-angle projection as in first, heights in the elevation and end view are equal, and depths in the plan are equal to corresponding widths in the end view.

Special arrangement of views

In exceptional circumstances it may not be possible to draw a view in the position required by the projection symbol. In this event the view may be placed in any convenient position, provided that the viewing direction is shown by an arrow. View Y in Exercise 4.9 is an example.

Scales

The proportions of an object must be preserved when a drawing of it is made. This can be achieved by **drawing to scale**. The scale which has been used must be stated on the drawing as a ratio; for example, ORIGINAL SCALE 1 : 2. BS 308 recommends that the scale should not be stated in words, such as full size, half size, etc.

The words 'original scale' are used because drawings are often reproduced on larger or smaller sheets and so to a different scale from that used when they were made. The words are a warning not to take measurements from the drawing with a rule.

When a drawing is made smaller than full size a **reduction scale** is used. Reduction scales are 1 : 2, 1 : 5, 1 : 10, 1 : 20, 1 : 50, 1 : 100, etc. An **enlargement scale** is used when the drawing is made larger than full size. Enlargement scales are 2 : 1, 5 : 1, 10 : 1, 20 : 1, 50 : 1, etc. When a drawing is made full size the scale is stated as 1 : 1.

The scale to be used for a drawing depends on the size of the drawing sheet and the size of the object to be drawn. The scale should be large enough to permit small details to be drawn and dimensioned easily. If this cannot be done, such details should be shown in a separate view to a larger scale, as illustrated in Figure 6.2.

Hidden detail

Most engineering parts have features which are internal and therefore cannot be seen in one or more of the three main views. Such **hidden detail** can be shown by using dashed lines instead of the continuous lines used for visible features.

In Figure 1.12 faces A and B are hidden in the end view and corner C is hidden in the plan. In Figure 1.13 face A is hidden in the elevation and the surface B of the hole is hidden in both elevation and end view. Note also that the centre line of the hole in Figure 1.13 is shown by a chain line, as is the line of symmetry of the part in the plan.

Everything which is visible in a view must be shown, as must all centre lines. But not all hidden detail need be shown in every view. In some cases, to show it all would be confusing instead of clarifying the drawing. The draughtsman or woman learns from experience how much to include. However, the rule mentioned above, that views should be chosen to ensure that as much information as possible is provided by visible lines, will help to prevent too much being included.

Information about hidden lines and centre lines is given in Figure 1.14, together with their correct representation in various situations.

Precedence of coincident lines

When two or more different types of line coincide, visible lines take precedence over hidden lines, and hidden lines take precedence over centre lines.

Self-assessment questions

1.5 What types of line are used on a drawing to represent visible edges and hidden edges of the object?

1.6 Thin lines on a drawing are half as thick as thick lines.   True or false?

1.7 Centre lines of holes are always shown as thin chain lines.   True or false?

1.8 All visible lines, hidden detail lines and centre lines must be shown on every view on a drawing. True or false?

Exercise 1.3

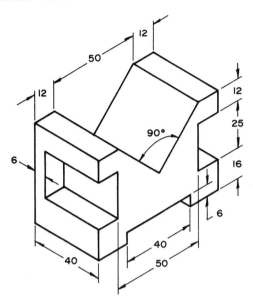

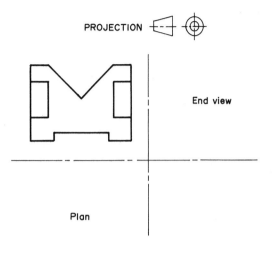

End view

Plan

Draw the given view and project two other views as shown with all hidden detail.

Exercise 1.4

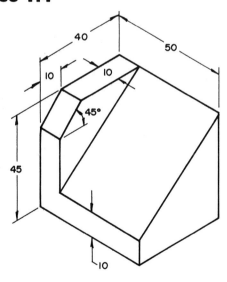

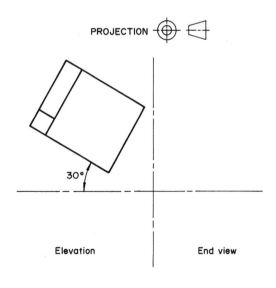

Elevation

End view

Draw the given view and project two other views as shown with all hidden detail.

13

Exercise 1.5

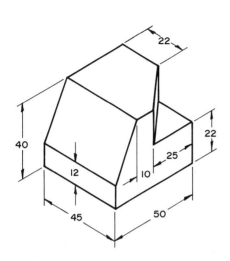

PROJECTION

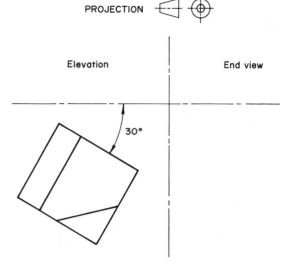

Elevation

End view

30°

Draw the given view and project two other views as shown.

Exercise 1.6

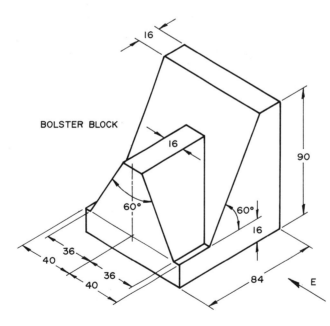

BOLSTER BLOCK

Draw the following views of the part using third angle projection:

(a) a front elevation looking in the direction of arrow E;

(b) a plan projected from the front elevation;

(c) an end view positioned on the left of the front elevation.

In each exercise draw the given views and project a third in the direction of the arrow. Show all hidden detail.

Exercise 1.7

PROJECTION

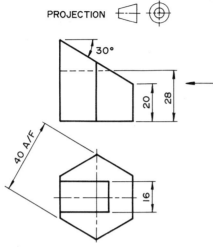

Exercise 1.8

PROJECTION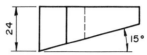

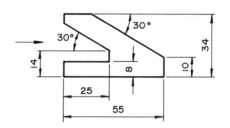

Exercise 1.9

PROJECTION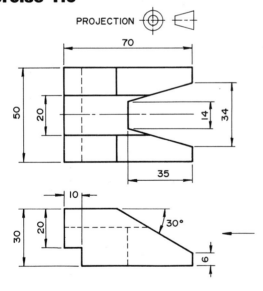

Exercise 1.10

PROJECTION

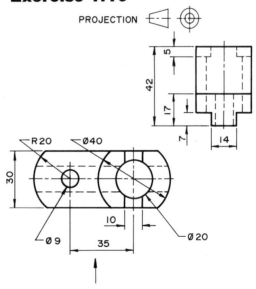

In each exercise draw the given views and project a third in the direction of the arrow
Show all hidden detail

In each exercise draw the two given views and project a third in the direction of the arrow.
Show all hidden detail.

Exercise 1.11

PROJECTION

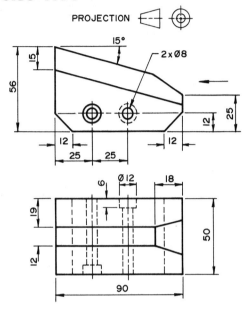

Exercise 1.12

PROJECTION

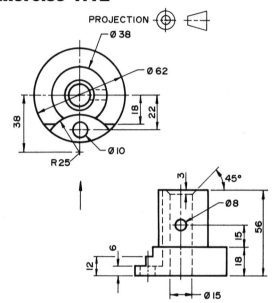

Exercise 1.13

PROJECTION

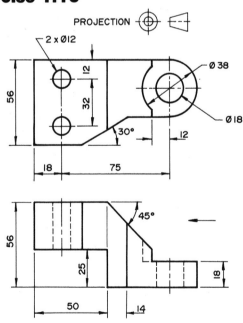

Exercise 1.14

PROJECTION

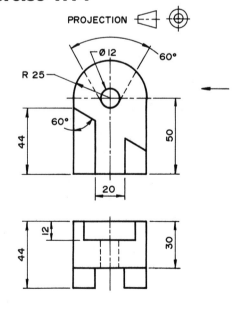

True lengths of lines

It is important to be able to find the **true length** of a line in some areas of engineering drawing, particularly in the development, or laying out in a plane, of sheet metal parts. It was stated earlier that the orthogonal projection of a line will show its true length only if the line is parallel to the plane on which it is projected. This is illustrated in Figure 1.15.

In Figure 1.15(a) the plan a^2b^2 of the line is parallel to the XY line. Hence the line is parallel to the vertical plane of projection, and its elevation on this plane is the true length (TL). Also the angle θ is the true inclination of the line to the horizontal plane.

The line AB in Figure 1.15(b) is parallel to the horizontal plane. Therefore, the plan a^2b^2 of the line shows its true length and the angle ϕ its true inclination to the vertical plane.

Figure 1.15(c) shows the elevation a^1b^1 of the line AB to be parallel to the auxiliary vertical plane, so the projection of the line on this plane gives its true length. Angle θ is the true inclination of the line to the horizontal plane.

In Figure 1.15(d) the line AB is parallel to both the vertical and the horizontal planes. Therefore its elevation a^1b^1 and its plan a^2b^2 are both true lengths. Its end view a^3b^3 is a point, from which it can be seen that a view of a line projected from a point view is true length.

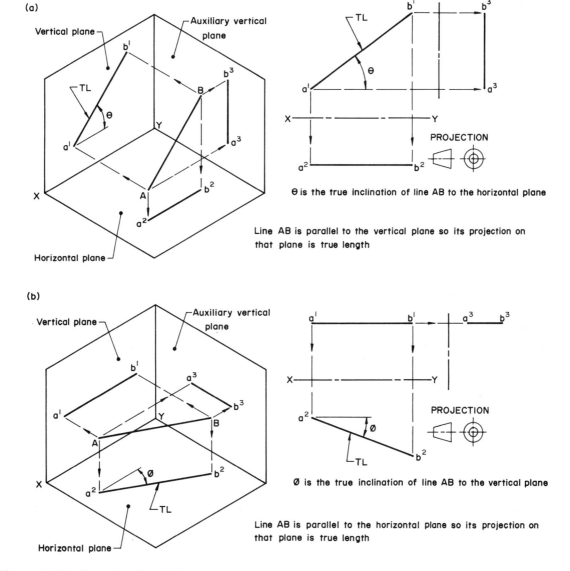

Figure 1.15 *True lengths of lines*

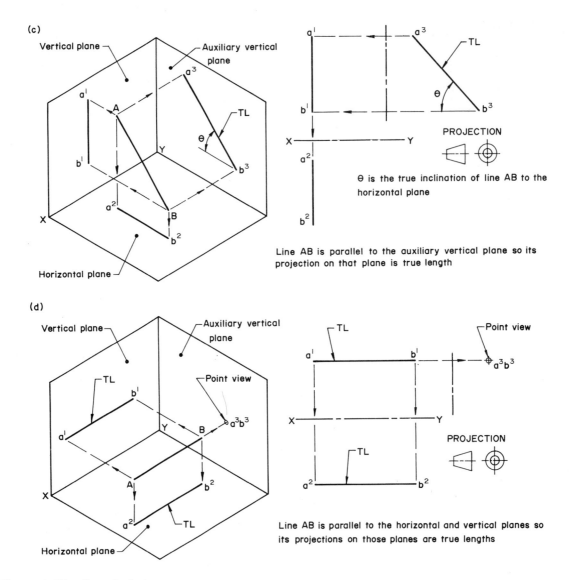

(c)

Vertical plane

Auxiliary vertical plane

TL

θ is the true inclination of line AB to the horizontal plane

Horizontal plane

Line AB is parallel to the auxiliary vertical plane so its projection on that plane is true length

PROJECTION

(d)

Vertical plane

Auxiliary vertical plane

TL

Point view

Horizontal plane

Line AB is parallel to the horizontal and vertical planes so its projections on those planes are true lengths

PROJECTION

Figure 1.15 *Concluded*

True lengths by revolution

When a line is not parallel to one of the normal planes of projection two methods may be used to find its true length. The line may be moved until it is parallel to a principal plane, or an auxiliary plane may be used which is parallel to the plan or elevation of the line. Figure 1.16 illustrates the first of these approaches, which is called the **revolution method**.

Figure 1.16(a) shows a line AB in space. The line is not parallel to any of the three planes of projection, so no projected view of the line is true length.

The same line AB is shown in Figure 1.16(b). The end A is kept stationary and the end B is revolved until the line is parallel to the vertical plane. Thus the

plan $a^2 b^2$ of the line will be parallel to the XY line. The plan b^2 of point B moves in a circle; the elevation b^1 moves in a straight line. Thus the point b^T is found and with it the true length. Angle θ is the true inclination of AB to the horizontal plane.

The lengths by triangulation

This method, illustrated in Figure 1.17, is essentially the same as the revolution method, but for some purposes it is more convenient. The plan length a^2b^2 is set off at right angles to the difference in vertical height of the ends of the line in elevation. The hypotenuse of the right-angled triangle so formed is

(a)

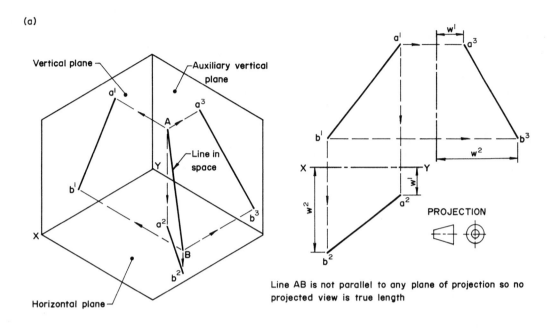

Line AB is not parallel to any plane of projection so no projected view is true length

(b)

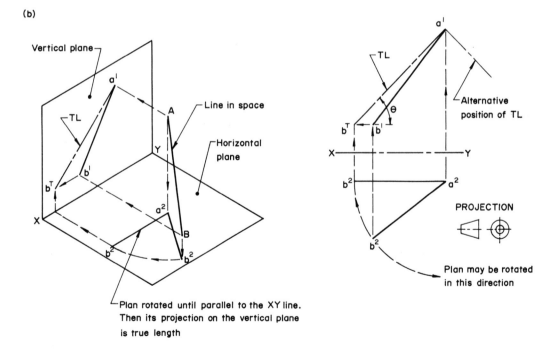

Plan rotated until parallel to the XY line. Then its projection on the vertical plane is true length

Plan may be rotated in this direction

Figure 1.16 *True length of a line by revolution*

the true length of the line, and angle θ is the true inclination to the horizontal plane.

True lengths by auxiliary views

The second approach to the true length of a line, using **auxiliary planes**, is shown in Figure 1.18. In Figure 1.18(a) the plan a^2b^2 is viewed at right angles in the direction of arrow S, and a view in this direction is drawn on an auxiliary plane whose ground line is X^1Y^1. This plane is parallel to a^2b^2, so the auxiliary elevation drawn on it will show the true length of AB. The heights h^1 and h^2 are the same in

both elevations and the angle θ is the true inclination of the line to the horizontal plane.

If the elevation a^1b^1 of the line is viewed at right angles in the direction of arrow T, as shown in Figure 1.18(b), an auxiliary plan projected on X^1Y^1, which is parallel to a^1b^1, will again show the true length of the line. The widths w^1 and w^2 are the same in both plans and may be transferred from the normal plan to the auxiliary plan. The angle ϕ is the true inclination of the line to the vertical plane.

In Figure 1.18 the views are drawn in first-angle projection. Third-angle projection may equally be used, as it can be with the revolution method.

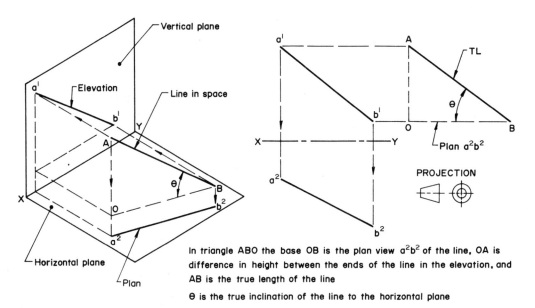

In triangle ABO the base OB is the plan view a^2b^2 of the line, OA is difference in height between the ends of the line in the elevation, and AB is the true length of the line

θ is the true inclination of the line to the horizontal plane

Figure 1.17 *True length of a line by triangulation*

(a)

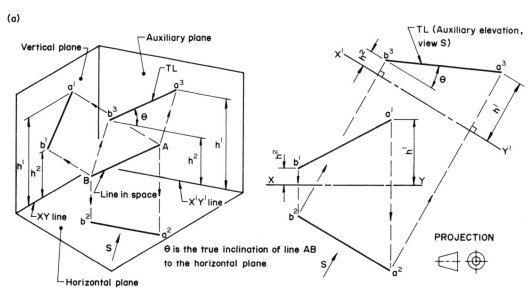

θ is the true inclination of line AB to the horizontal plane

Figure 1.18 *True length of a line by auxiliary views*

(b)

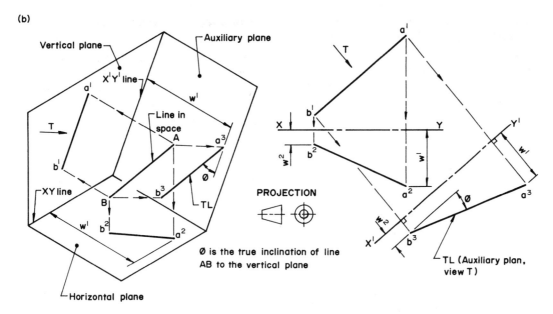

Ø is the true inclination of line
AB to the vertical plane

PROJECTION

TL (Auxiliary plan,
view T)

The new ground line X'Y' for the auxiliary view must be parallel to
a principal view of the line

Figure 1.18   *Concluded*

Self-assessment questions

1.9   In an orthographic view the projection of a line can be longer or shorter than the line itself. True or false?

1.10   The elevation of a line is 10 long and is parallel to the XY line. Does
(a) the elevation,
(b) the plan, or
(c) an end view
show the true length of the line?

1.11   To find the true inclination of a line to the horizontal plane, does
(a) the elevation,
(b) the plan, or
(c) an end view
have to be a true length?

True shapes of surfaces

Three methods are available for finding the **true shape** of a surface: by finding the true lengths of the sides; by revolution; and by using auxiliary views.

True shapes by use of true lengths

Figure 1.19 shows a triangle ABC in space in the first quadrant. The true lengths of sides AB and BC have been found by revolution, and of side AC by

21

triangulation. Using these true lengths the true shape (TS) of the triangle has been constructed.

This method may be used to find the true shape of any plane surface, of whatever shape, provided that it is first divided into triangles. The true shapes of the triangles are then laid out in order, side by side.

True shapes by revolution

A truncated pyramid positioned in the first quadrant is shown in Figure 1.20. The true shape of the sloping face BCDE may be found by revolving it about the side BC until the face is parallel to the horizontal plane. Then its projection on that plane will be true shape.

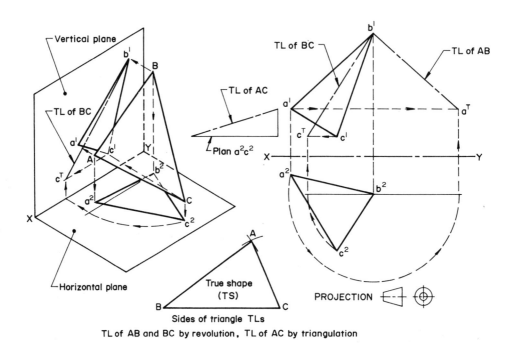

Figure 1.19 *True shape by use of true lengths*

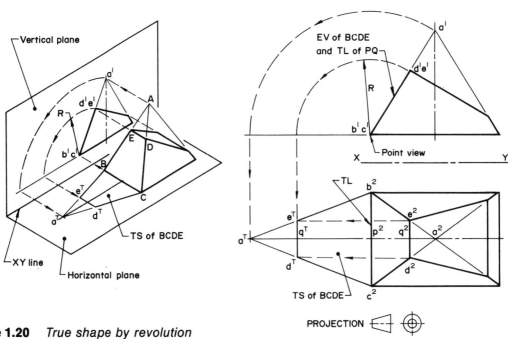

Figure 1.20 *True shape by revolution*

22

If it is more convenient, a face may be revolved until parallel to the vertical plane or an auxiliary vertical plane. For example, in Figure 1.20 face BCDE could be revolved until it was parallel to one of the auxiliary vertical planes, when its projection on that plane would be true shape.

To use the revolution method, the face for which the true shape is required must appear in edge view (EV) and one side must be a point view. The point view is the hinge about which the face must be revolved.

True shapes by auxiliary views

It was stated earlier in this chapter that the true shape of a surface can be projected onto a plane which is parallel to the surface. In Figure 1.21(a) the

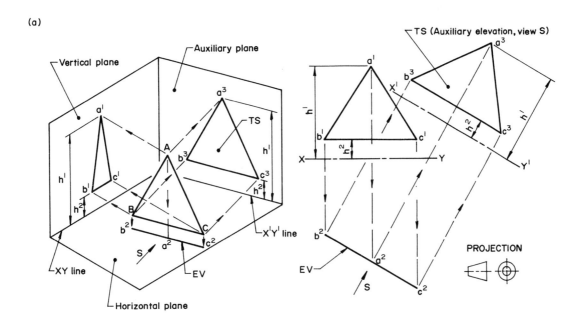

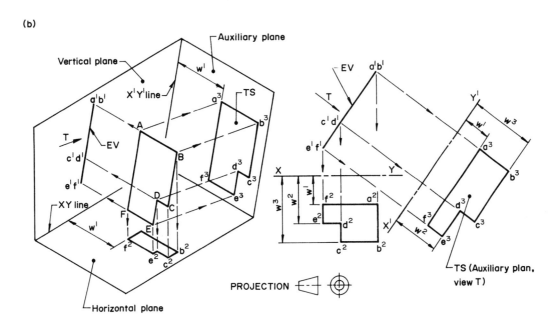

The X'Y' line must always be parallel to the edge view of the plane

Figure 1.21   *True shape by auxiliary views*

23

triangle is not parallel to either the vertical or the horizontal plane so its projections on those planes will not be true shape.

However, it is normal to the horizontal plane, so its projection on that plane is an edge view. If an auxiliary plane is added which is parallel to the edge view, then the projection on that plane will be true shape. In the orthographic views this means that the new ground line X^1Y^1 must be parallel to the edge view. In Figure 1.21(a) the heights h^1 and h^2 can be transferred to the auxiliary view to enable it to be completed.

In Figure 1.21(b) the surface ABCDEF is perpendicular to the vertical plane so its projection on that plane is an edge view. When the surface is projected onto an auxiliary plane which is parallel to the edge view, that projection will be true shape. Note that the new ground line X^1Y^1 is again parallel to the edge view, and that depths w^1, w^2 and w^3 are transferred to the auxiliary view.

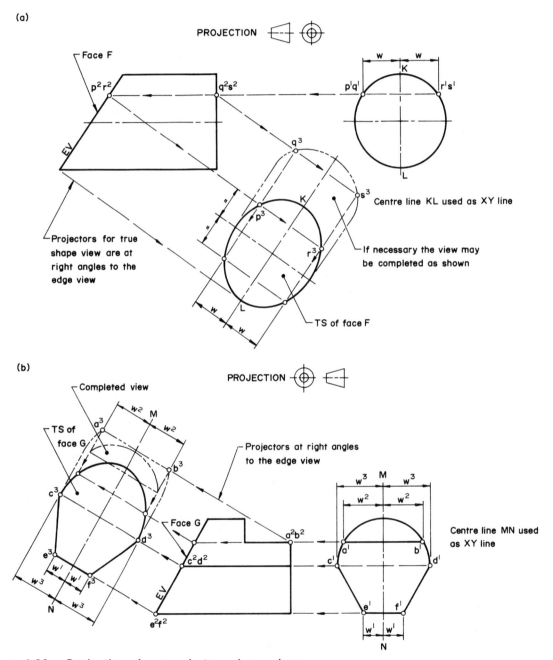

Figure 1.22   *Projection of curves in true shape views*

Projection of curves in true shape views

Curves may be projected into auxiliary views which show the true shape of a surface by projecting a series of points on the curve in the same way that points at the ends of straight lines are projected. Figure 1.22 illustrates the procedure.

In Figure 1.22(a) widths w from the circular view are transferred to the true shape view along projectors which are perpendicular to the edge view of face F. Note that, for convenience, the centre lines

of the circular view and true shape view are used as XY lines. If the complete auxiliary view is required then it may be projected in the same way.

When sufficient points have been projected, the curves may be lined in using French curves. If too many points are projected the work becomes tedious and there is no increase in accuracy.

Figure 1.22(b) shows an example in third-angle projection. The method is identical to that used for first-angle, but now the true shape view is drawn adjacent to the edge view of face G.

Exercise 1.15

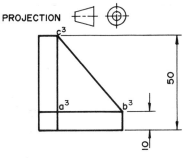

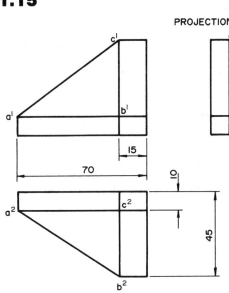

Draw the given views of the angle block and indicate the true lengths of the lines AB, AC and BC.

Dimension the true inclination of AB to the vertical plane, BC to the auxiliary vertical plane and AC to the horizontal plane.

Exercise 1.16

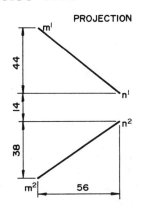

Using the revolution method determine the true length of the line MN and its true angles to the vertical and the horizontal planes.

Exercise 1.17

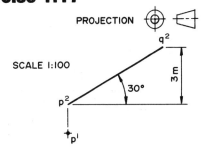

SCALE 1:100

Complete the front view of the water main PQ which slopes downwards at a true angle of 15° from P. Indicate the true length.

Exercise 1.18

PROJECTION

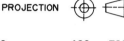

DIMENSIONS IN m
SCALE 1:100

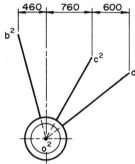

Indicate the true lengths of the tripod members OA, OB and OC.

Exercise 1.19

PROJECTION

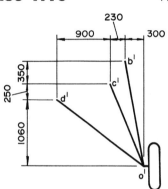

Draw the views of the light aircraft undercarriage to a scale of 1 : 20 and find the true lengths of the struts OB, OC and OD by the revolution method.

Exercise 1.20

PROJECTION

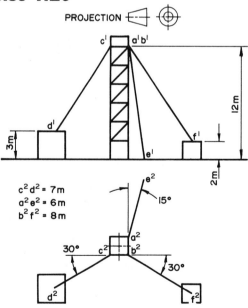

$c^2 d^2 = 7m$
$a^2 e^2 = 6m$
$b^2 f^2 = 8m$

The tower for microwave transmitter dishes is stabilized by three guy wires AE, BF and CD. Using auxiliary views find the true lengths of the wires and their true inclinations to the horizontal plane. Scale 1 : 200.

Exercise 1.21

PROJECTION

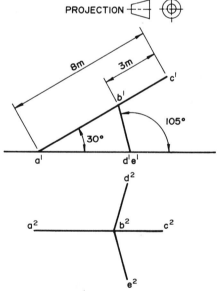

The guided missile launching rail AC is supported at 30° to the horizontal by the legs BD and BE. Find the true lengths of the legs if the angle between them is 90° and the true angles they make with the horizontal. Project the plan view. Scale 1 : 100.

Exercise 1.22

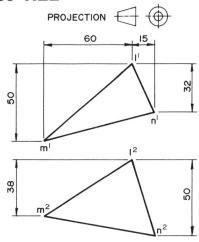

Find the true lengths of the sides of the triangle LMN and from them construct its true shape and calculate its area.

Exercise 1.23

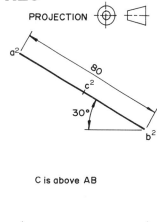

C is above AB

ABC is a 45°–45°–90° triangle. Find its true shape and complete the elevation.

Exercise 1.24

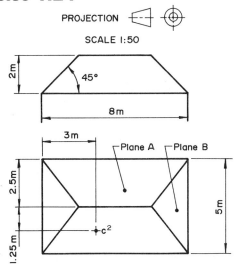

Find by revolution the true shapes of roof planes A and B and calculate the total area of the roof. Show in both views a hole centred at C, cut in the roof for a 1 m square chimney.

Exercise 1.25

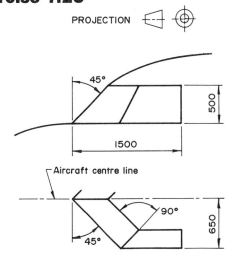

Using auxiliary views find the true shapes of the aircraft window panels. Scale 1 : 10.

Exercise 1.26

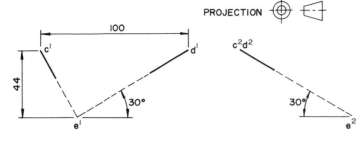

The corner E is to be rounded off using a 40 radius. Find the points of tangency on the lines CE and DE.

Exercise 1.27

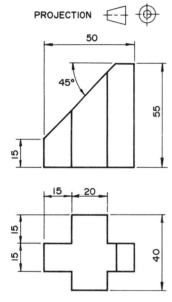

Draw the given views and project the true shape of the inclined face.

Exercise 1.28

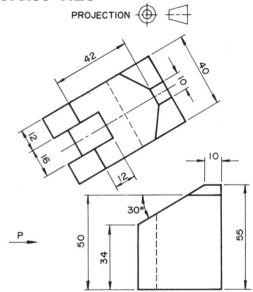

Draw the given views and project an end view on arrow P.

Exercise 1.29

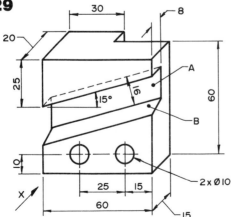

Using first angle projection draw an elevation in the direction of arrow X, a plan and an auxiliary view to show the true angle of 120° between the surfaces A and B.

Show all hidden detail in the plan and elevation.

28

Exercise 1.30

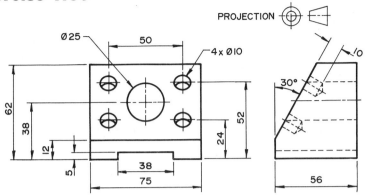

PROJECTION

From the given views project a partial auxiliary view to show the true shape of the sloping face.

Exercise 1.31

PROJECTION

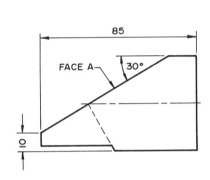 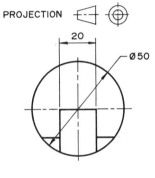

From the given views project a complete auxiliary view which shows the true shape of face A.

Exercise 1.32

PROJECTION

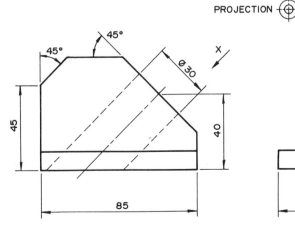 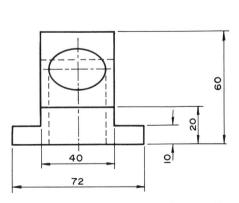

Draw the given views and project a complete view in the direction of arrow X.

29

Self-assessment questions

1.12 Any view projected from the point view of a line will show the true length of the line. True or false?

1.13 To find the true shape of a surface by using an auxiliary view, the projectors for the auxiliary view must be at 30°, 60° or 45° to the XY line. True or false?

1.14 The XY line for a new view is always perpendicular to the projectors for the new view. True or false?

Post test 1

1 (a) The principal views in orthographic projection are those on the vertical and auxiliary vertical planes. True or false?
(b) Hidden detail is represented by thin dashed lines. True or false?
(c) A scale of 5 : 1 is an enlargement scale. True or false?
(d) If a line is parallel to the horizontal plane the plan will show its true length and true inclination to the vertical plane. True or false?

2 Decide whether the given elevation and end view are drawn in first- or third-angle projection, draw them and add the plan

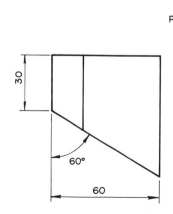
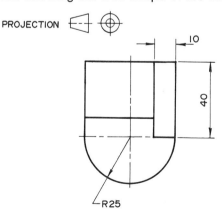

3 Draw the given views and project a view showing the true shape of the inclined face only

PROJECTION

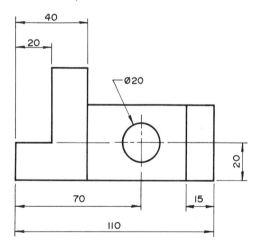
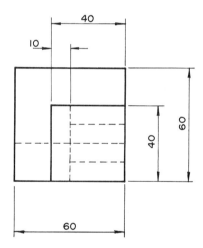

Sections and sectional views

Principal learning objective

The student uses the conventions in British Standards to produce sectioned views.

Introduction

Simple objects with little interior detail can be represented satisfactorily in orthographic projection by views of the exterior, hidden detail lines being used to show the interior construction. When the interior detail is extensive, the hidden detail lines may be confusing and difficult to interpret correctly. In such cases the draughtsman or woman imagines that the object is cut by a plane, as in Figure 2.1. The part of the object between the eye and the plane is then assumed to be removed. This exposes the interior detail which can then be shown by continuous lines instead of the dashed hidden detail lines. The resulting view is a **sectional view** or a **section**.

Cutting planes

Cutting planes are represented by thin chain lines, thickened at the ends and at any changes of direction. The direction in which they are to be viewed is given by arrows which touch the middle of the thickened ends. Letters on the arrows identify the cutting plane and are repeated below the resulting sectional view. These points are illustrated in Figure 2.1.

The position of the cutting plane is selected to show the interior of the object to the best advantage. When its position is obvious, for example when it coincides with the centre line of a symmetrical part, it is often omitted.

Full sections

When the cutting plane passes completely through the part, as in Figure 2.1, the resulting view is called a

full section. Note that all visible edges behind the cutting plane must be shown or the view will be incomplete, as shown in Figure 2.2. Such a view is meaningless. Hidden detail lines however, are not shown on a sectional view unless they are needed to describe the part completely.

Note that other views on the drawing show the complete part, unless they too are sectional views.

Section lines or hatching

A sectional view is distinguished from an outside view by **section lines** or **hatching** drawn on the cut surfaces produced by the section plane, see Figure 2.1. Thin lines inclined at 45° are used for hatching. They are equally spaced by eye, the spacing being not less than about 4 mm. The larger the area to be hatched the wider the spacing may be, up to a maximum of about 10 mm. For very large areas the hatching may be limited to a zone following the contour of the hatched area, as shown in Figure 2.3. Hatching may be omitted from a sectional view if the meaning of the drawing is clear without it. This has been done in Figure 2.7(a) and Figure 2.10.

It is important to the appearance of the finished drawing that hatching is drawn carefully. The spacing must be consistent, not too close, and the lines must touch the outlines of the section and be thin.

When a single component is sectioned, as in Figure 2.4(a), the slope and spacing of the hatching must be the same throughout the view. On an assembly drawing the slope of the hatching must be reversed on adjacent parts, as shown in Figure 2.4(b). For each component the slope and spacing of the hatching must be the same on all views on the drawing, as in Figure 2.11. If more than two parts are in contact the angle of the hatching may be changed from 45°, or the spacing may be varied to avoid the impression that it is crossing outlines. Hatching never crosses outlines.

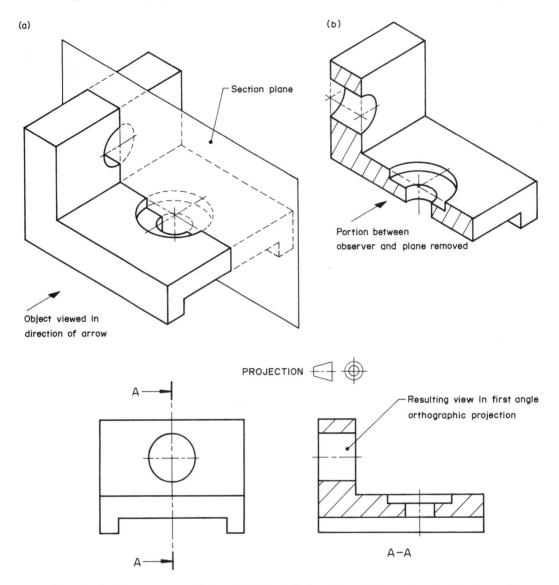

The rules for the arrangement of views in first and third angle projection apply when drawing sections

Figure 2.1 *Interior shown by a section*

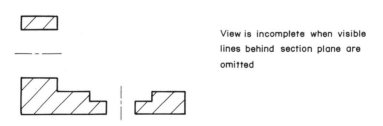

View is incomplete when visible lines behind section plane are omitted

Figure 2.2 *Background lines*

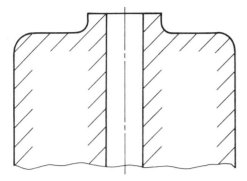

Figure 2.3 *Hatching of large areas*

(a)

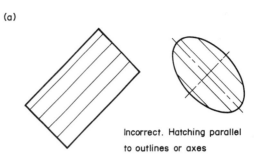

Incorrect. Hatching parallel
to outlines or axes

(b)

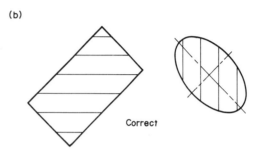

Correct

Figure 2.5 *Alteration of the slope of hatching*

(a)

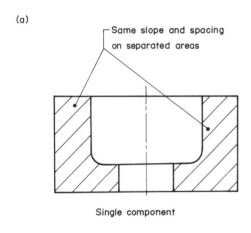

Same slope and spacing
on separated areas

Single component

(a)

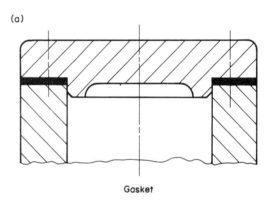

Gasket

(b)

Slope of hatching
reversed

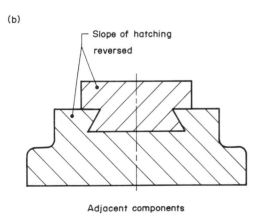

Adjacent components

Figure 2.4 *Section lining on adjacent components*

(b)

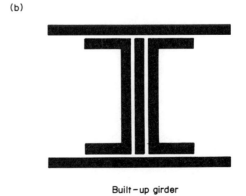

Built-up girder

Figure 2.6 *Sectioning of thin parts*

Alteration of the slope of hatching

Cases sometimes occur where the outlines or axes of a sectional view slope at 45°. If the hatching is also drawn at 45° a misleading impression is given, as in Figure 2.5(a). The hatching should therefore be drawn at some other easily obtained angle, such as 30° or 60°, or drawn horizontally or vertically as in Figure 2.5(b).

Thin material in section

When sheet metal parts, gaskets and other parts made from thin material appear on sectional views, the areas involved may be too thin for hatching. Instead of drawing the material thickness out of scale, the section may be filled in as shown in Figure 2.6. If two or more thin details are joined in a sectional view on an assembly drawing, spaces should be left between them or the drawing will be difficult to read. This has been done in the built-up girder in Figure 2.6(b).

Self-assessment questions

2.1 What type of line is used to represent a cutting plane?

2.2 On a full sectional view all visible lines behind the cutting plane may be shown or they may be omitted.   True or false?

2.3 Hatching must always be shown on a sectional view.   True or false?

2.4 Hatching is generally drawn at 45° to the horizontal.   True or false?

Exercise 2.1

Exercise 2.2

PROJECTION ◁ ⊕

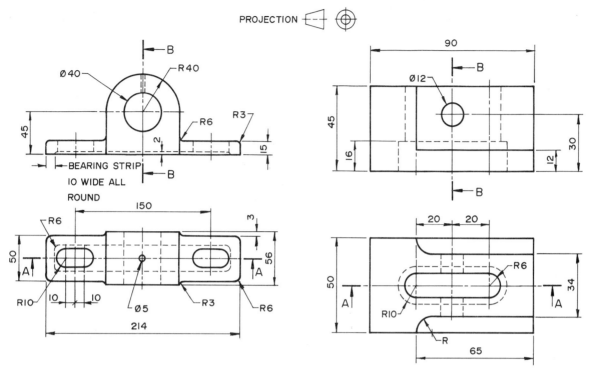

In each exercise draw the given plan and add sections A–A and B–B.

Exercise 2.3

Exercise 2.4

PROJECTION ⊕ ◁

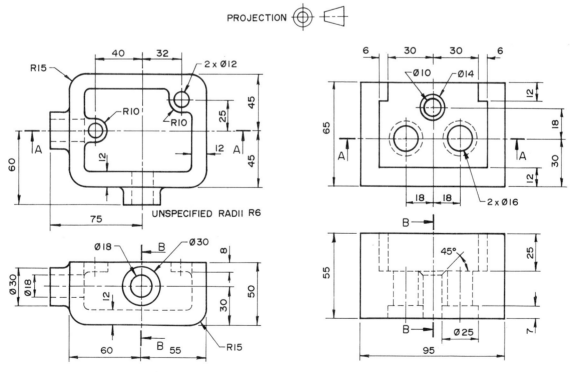

In each exercise draw the given plan and add sections A–A and B–B.

Half sections

Symmetrical parts may be shown to advantage by **half sections**, that is with one half drawn in section and the other shown as an outside view. Thus two views are combined in one with a consequent saving of draughting time and space. Half sections have the disadvantage that the dimensioning of the hidden features may be difficult without using hidden detail lines, and for this reason they are more often used for assembly drawings. Few dimensions, if any, are needed on assembly drawings, and hidden detail can be omitted from the unsectioned half of the view. Examples of half sections are given in Figure 2.7. Note that the halves of a half section are separated by a centre line and not by an outline.

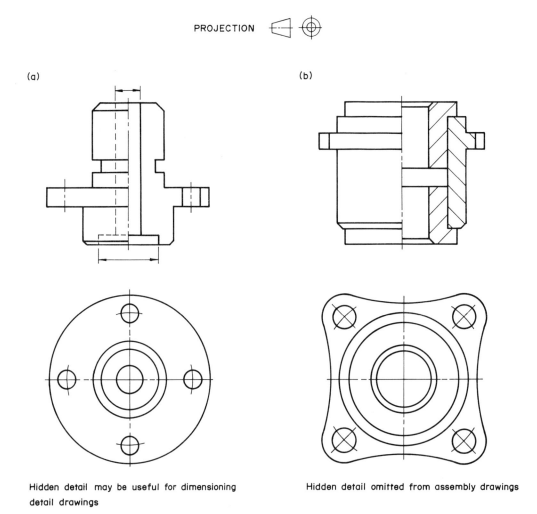

PROJECTION

(a)

(b)

Hidden detail may be useful for dimensioning detail drawings

Hidden detail omitted from assembly drawings

Figure 2.7 *Half sections*

In both exercises draw the given circular view and change the other to a half section.

Exercise 2.5

PROJECTION

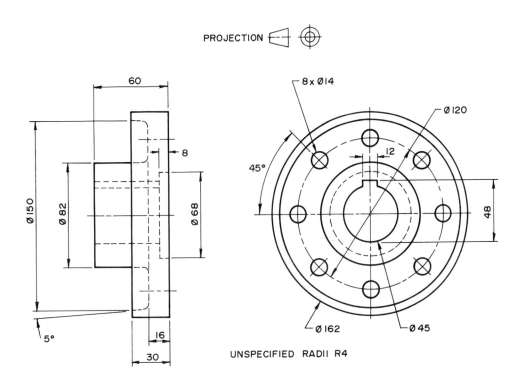

8 x Ø14

Ø 120

60

8

45°

12

Ø 150

Ø 82

Ø 68

48

5°

16

30

Ø 162

Ø 45

UNSPECIFIED RADII R4

Exercise 2.6

PROJECTION

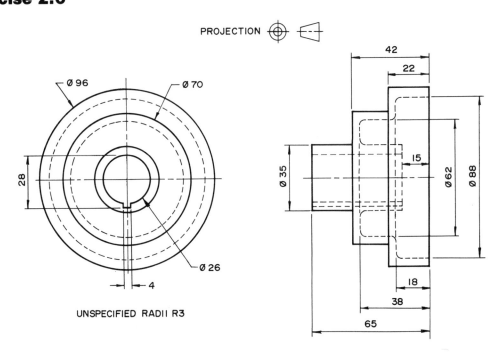

Ø 96

Ø 70

42

22

28

Ø 35

15

Ø 62

Ø 88

Ø 26

4

18

UNSPECIFIED RADII R3

38

65

Exercise 2.7

PROJECTION

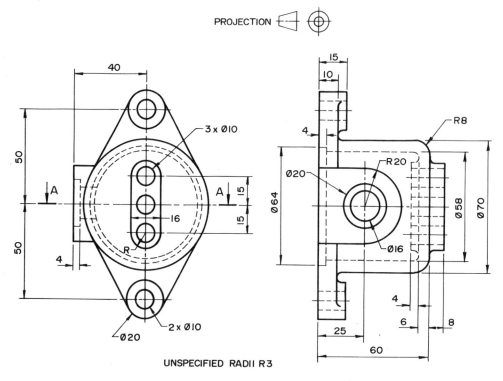

UNSPECIFIED RADII R3

Draw the left-hand view, change the right-hand view to a half section and add sectional plan AA.

Exercise 2.8

PROJECTION

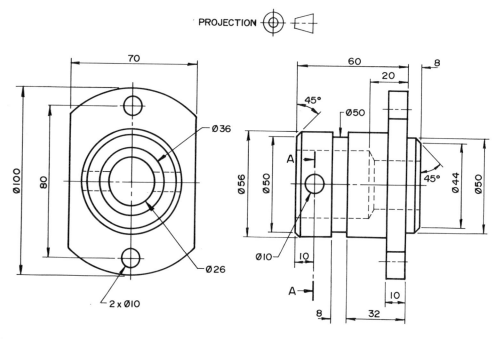

Change the right-hand view to a half section with the top half in section and the left-hand view to a half section on AA.

Local or broken-out sections

Occasionally a part of a view in section is all that is needed to show the internal details of an object. Sometimes a whole view cannot be sectioned because it would conceal external features which must be shown. In such cases a **local** or **broken-out section** is useful. The cutting plane is imagined to pass part way through the object and the piece in front of the plane is then assumed to be broken away, leaving an irregular boundary. Note that this boundary is a thin line.

In Figure 2.8(a) the adjusting screw is solid except for the two holes in the head, and these are shown by a local section through the head. In Figure 2.8(b) a complete half section would have removed half of the keyway, which might have been misleading. All of the keyway is shown by stopping the half section with a break line. Other examples of local sections are shown in Figure 2.15.

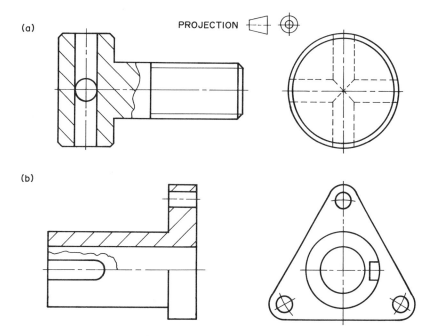

Figure 2.8   *Local or broken-out sections*

Self-assessment questions

2.5   Do cutting planes have always to pass completely through an object?

2.6   State two advantages of using half sections.

2.7   Is the irregular boundary of a local section a thin or a thick line?

Revolved sections

These sections are drawn directly on an outside view and are used to show the local cross-section of an arm, rib or similar feature. The cutting plane is assumed to pass through the part at right angles to the axis of the feature and is then revolved into the plane of the paper. Illustrations of revolved sections are shown in Figure 2.9. Note that the outline of the section is a thin line, and any outlines on the main view which are covered by the section are not shown.

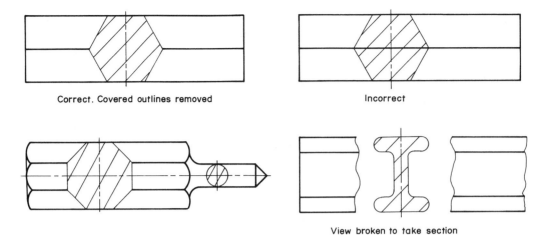

Correct. Covered outlines removed

Incorrect

View broken to take section

Outlines of revolved sections are thin lines

Figure 2.9 *Revolved sections*

Removed sections

These are similar to revolved sections but instead of being drawn on the outside view they are removed to another part of the drawing. Their outlines are the normal thick line. They are used in place of revolved sections when lack of space prevents the section being drawn or dimensioned on the outside view. To improve their clarity they are sometimes drawn to a larger scale than the remainder of the drawing. They are not subject to the strict rules of projection regarding their position on the drawing. If they are numerous they are occasionally grouped together on a separate sheet. For these reasons it is essential that the position of the cutting plane is shown and the section is labelled to relate it to the cutting plane.

Removed sections are illustrated in Figure 2.10. Note that features behind the cutting plane are not shown. These sections are usually fairly simple, and hatching may be omitted without affecting the readability of the drawing.

Part or scrap sections

It is necessary occasionally to show part of an object in section to describe a small detail of the construction, but the views may be unsuitable for the use of a broken-out section. In such a case a **part** or **scrap section** may be used. This is similar to a broken-out section but is drawn away from the outside view. The part section, where it is broken off, has an irregular boundary, which is thin. Frequently

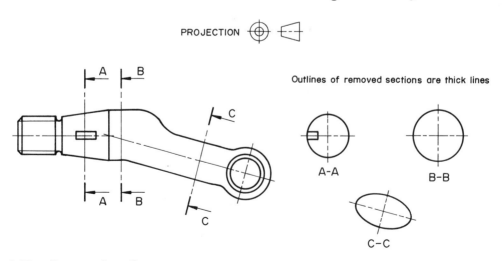

PROJECTION

Outlines of removed sections are thick lines

A-A

B-B

C-C

Figure 2.10 *Removed sections*

background features are not shown. Examples are given in Figure 2.11. Part sections, like removed sections, are often drawn in any convenient position on the drawing, and therefore should be given a title.

Offset sections

On a full section the cutting plane need not pass straight through a part but may be offset as required to include features which are not in a straight line. Figure 2.12 is an example. The position of the cutting plane must be shown and the resulting sectional view should carry a title. Where the cutting plane changes direction thick dashes are used and these dashes touch to form a right angle. Note that at the places where the plane changes direction on the sectional view, no lines are shown.

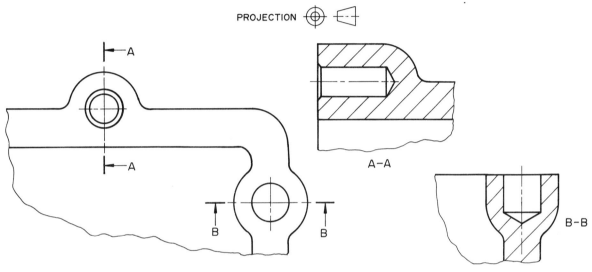

Figure 2.11 *Part or scrap sections*

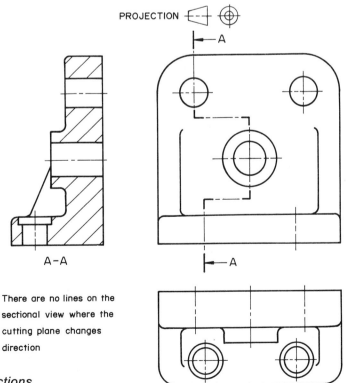

Figure 2.12 *Off-set sections*

There are no lines on the sectional view where the cutting plane changes direction

41

Exercise 2.9

PROJECTION ⊕ ◁

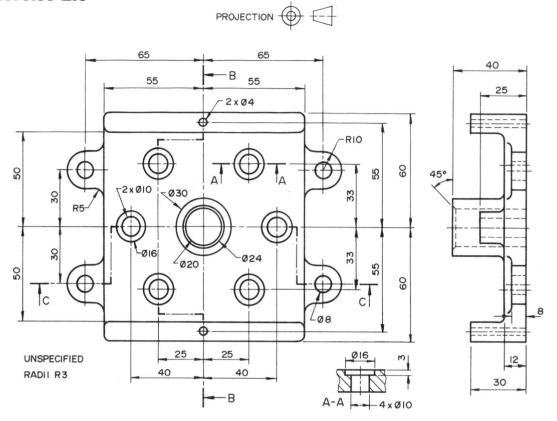

UNSPECIFIED RADII R3

Draw the left-hand view, change the right-hand view to section BB and add a sectional view on CC.

Exercise 2.10

PROJECTION ▷ ⊕

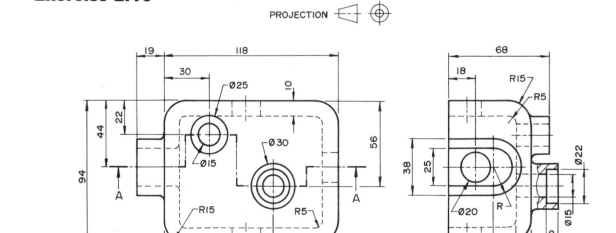

UNSPECIFIED RADII R3

Draw the left-hand view and project from it a sectional view on AA.

Aligned sections

These are used when sectioning parts whose features lie on radial lines, such as the part in Figure 2.13. The cutting plane generally coincides with a main centre line and one or more radial centre lines.

The section is drawn with the features on the radial centre lines aligned with the main centre line. By this means awkward projections are avoided and the features through which the section plane passes are shown at their true distances from the centre of the part. As with offset sections, where the cutting plane changes direction it is drawn as a thick line, and these thick lines touch. Other examples of aligned sections are shown in Figure 2.14(d) and (e).

Alignment is used also on outside views to avoid tedious projections. For example, if the section of the bell crank lever in Figure 2.13 were drawn as an outside view, it could still be aligned to make the drawing easier.

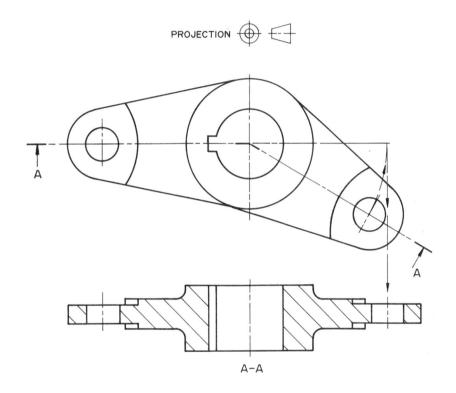

The arrows are always at right angles to the section plane

Figure 2.13 *Aligned sections*

Self-assessment questions

2.8   What type of line is used for the outlines of a revolved section and a removed section?

2.9   On a part section, do all features behind the cutting plane have to be shown?

2.10   Is it essential to show the position of the cutting plane when drawing an offset section?

Exercise 2.11

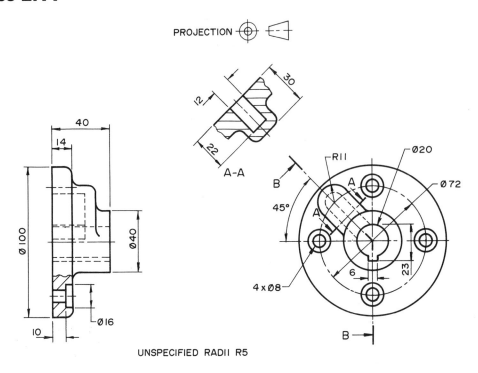

PROJECTION

A-A

UNSPECIFIED RADII R5

Draw the right-hand view and replace the left-hand view with section BB.

Exercise 2.12

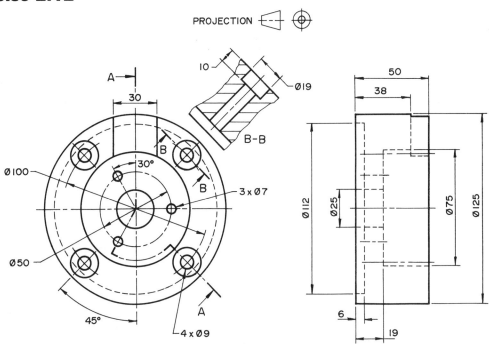

PROJECTION

B-B

Draw the left-hand view and replace the right-hand view with section AA.

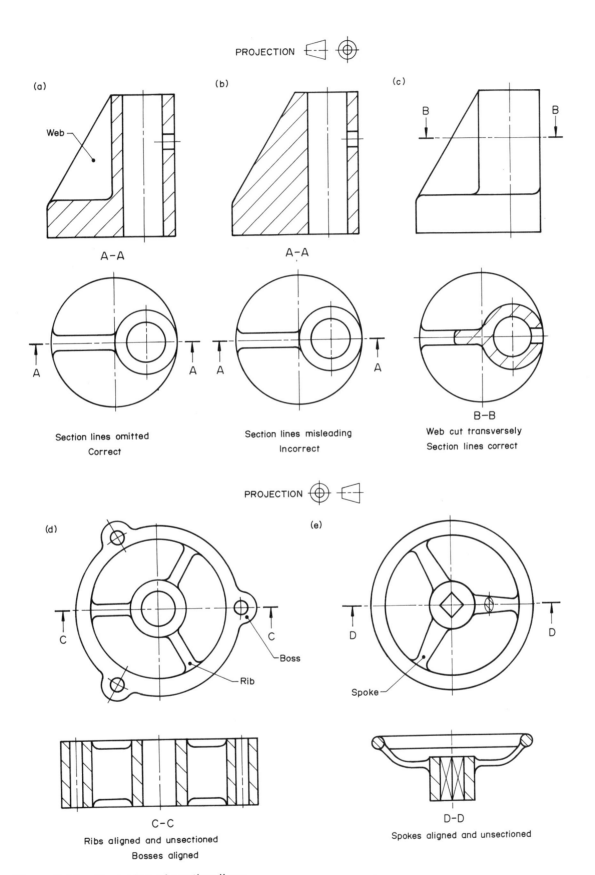

PROJECTION

(a)

Web

A–A

Section lines omitted
Correct

(b)

A–A

Section lines misleading
Incorrect

(c)

B   B

B–B
Web cut transversely
Section lines correct

PROJECTION

(d)

C   C

Boss

Rib

C–C
Ribs aligned and unsectioned
Bosses aligned

(e)

D   D

Spoke

D–D
Spokes aligned and unsectioned

Figure 2.14   *Omission of section lines*

Features of a part from which hatching is omitted

Whenever hatching on a sectional view would result in a misleading effect it should be omitted. For example, consider the casting shown in Figure 2.14(a), (b) and (c). Section A–A passes longitudinally through a triangular web whose purpose is to strengthen the part. If the section lines are drawn on the web, as at (b), the impression of a partly conical shape is given. This impression is incorrect so the hatching is omitted and (a) is the correct representation. Note that the web is bounded by an outline where it runs into the base and circular boss. These outlines should not be left out or replaced by hidden detail lines. If the web is cut transversely by the cutting plane, as at (c), it must always carry the hatching.

Further examples of the omission of hatching are given in Figure 2.14(d) and (e). Here the ribs and spokes are not hatched and to avoid tedious projections they are aligned on to the plane of the section. Gear teeth are treated in a similar way.

Parts which are not sectioned

In addition to the examples of omission of hatching given above, some parts are not sectioned on assembly drawings when they are cut lengthwise by the section plane. These parts include nuts, bolts, screws, studs, rivets, solid shafts and small solid cylindrical parts, keys, cotters, split and taper pins and balls and rollers in bearings.

These parts are not sectioned because they have no internal features, and also because they are more easily recognised by their outside views than by a section. Figure 2.15 shows two small assemblies to which this principle has been applied. However, if these parts are cut transversely, resulting in a circular sectional view, they are hatched in the normal way.

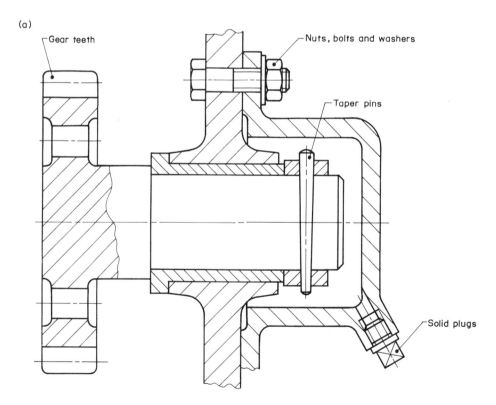

(a)

Gear teeth

Nuts, bolts and washers

Taper pins

Solid plugs

Figure 2.15 *Details not sectioned on assemblies*

(b)

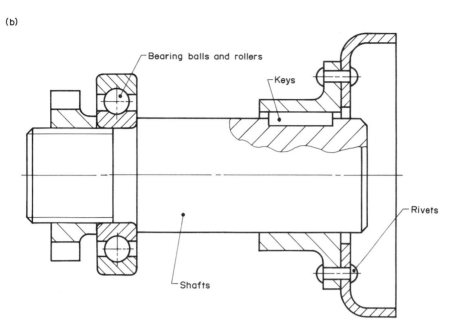

Figure 2.15   *Concluded*

Exercise 2.13

PROJECTION

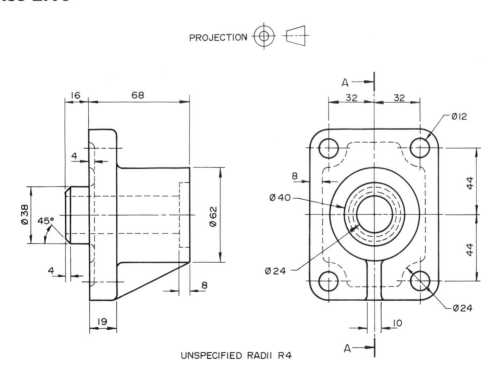

UNSPECIFIED RADII R4

Draw the right-hand view and replace the left-hand view with section AA.

Exercise 2.14

PROJECTION

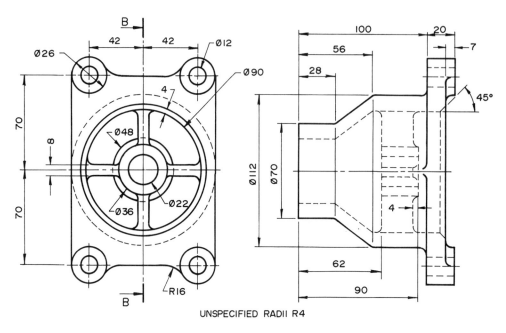

UNSPECIFIED RADII R4

Draw the left-hand view and replace the right-hand view with section BB.

Self-assessment questions

2.11 A sectional view is always an auxiliary view.   True or false?

2.12 On a sectional view the boundary between an unsectioned web and the remainder of the view is hidden detail line.   True or false?

2.13 When is a shaft completely hatched on a sectional view?

2.14 What advantages result from using an aligned section?

Post test 2

1 (a) Hidden detail must always be shown on a sectional view.   True or false?
(b) Hatching is always shown with thin lines.   True or false?
(c) State when a section may be filled in instead of being hatched.
(d) Give the name of the type of section which is drawn directly on an outside view.
(e) A section plane must always be straight.   True or false?
(f) Name four parts which are not sectioned when cut lengthwise by a section plane.

2 Draw the right-hand view only, changing it to a half section.

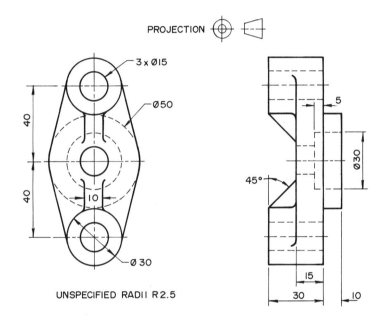

PROJECTION

3 x Ø15

Ø50

40

40

10

Ø 30

UNSPECIFIED RADII R 2.5

5

Ø30

45°

15

30

10

3 Draw removed sections AA, BB and CC only.

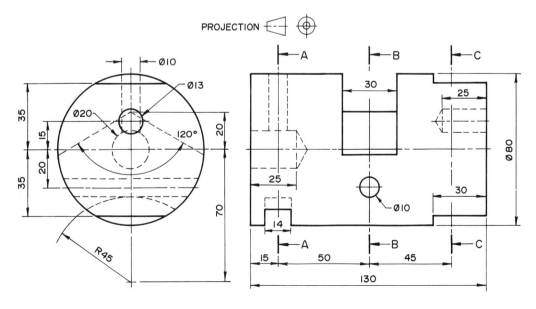

PROJECTION

Ø10

Ø13

Ø20

120°

35

15

20

35

20

70

R45

A

B

C

30

25

25

Ø80

25

Ø10

30

14

A

B

C

15

50

45

130

Dimensioning

Principal learning objective

The student uses the conventions shown in British Standards to produce dimensioned working drawings.

Introduction

The study of dimensioning falls into two parts. First, the technique of dimensioning, that is, the correct drawing, lettering and positioning of the dimensions on the drawing, according to accepted rules and conventions. Secondly, dimensioning requires that dimensions are selected which will ensure that the finished part will function correctly, and enable it to be made without any sizes having to be calculated from the given dimensions. The basic principles for the selection of dimensions are given later in this chapter, but a full understanding of this side of dimensioning requires experience in making drawings, and in the workshop processes by which the part is made.

Dimensioning technique

Projection and dimension lines

Dimensions should be placed outside the outlines of the view wherever possible. This is achieved by drawing **projection lines** from points or lines on the view and placing a **dimension line** between them, as shown in Figure 3.1.

Projection and dimension lines are thin and continuous and dimension lines end in arrowheads. Arrowheads should be narrow, 3 mm to 5 mm long and filled in. Figure 3.1 shows that projection lines start just clear of the outline of the feature and extend a little way beyond the dimension line. Projection lines are normally at right angles to the dimension line.

Centre lines and their extensions, and outlines and their extensions must not be used as dimension lines. They may, however, be used as projection lines.

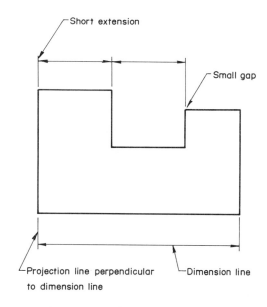

Projection and dimension lines are thin

Arrowheads are narrow and 3 to 5mm long

Figure 3.1   *Projection and dimension lines*

Preferably dimension and projection lines should not cross. Where they must cross, as a general rule they are not broken at the crossing point. Breaks are shown only for clarity and then only in the projection line. Breaks are never shown in outlines.

Figure 3.2 illustrates some recommended uses of dimension lines. Note that to make the drawing easy to read the dimension lines are well spaced. Larger dimensions are placed outside smaller dimensions, to avoid dimension lines and projection lines crossing.

Where partial views of symmetrical parts are drawn, as in Figure 3.3, the parts of the dimension lines should extend a short distance beyond the axis of symmetry. Only one arrowhead is shown, but the full value of the dimension must be given. See also Exercise 6.6.

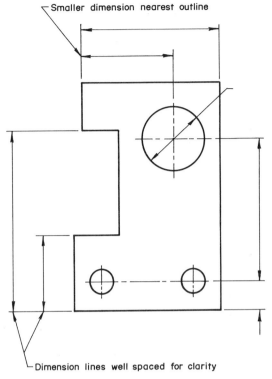

Figure 3.2 *Examples of dimension lines*

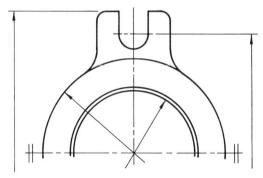

Figure 3.3 *Dimension lines on a partial view of a symmetrical part*

Dimensions

Linear dimensions on drawings are usually in millimetres, for which the unit symbol is 'mm'. To avoid having to show the unit symbol with each dimension, the drawing generally carries a note such as 'Dimensions are in millimetres'.

Dimensions should be shown using the least number of significant figures, for example, 50 not 50.0. For ways to show toleranced dimensions see

Figure 3.29. Dimensions which are less than one should have a zero before the decimal point, for example 0.75. The decimal point should be bold and placed on the base line of the figures.

Angular dimensions should be shown in degrees, degrees and minutes, or degrees, minutes and seconds, depending on how much accuracy is needed. When an angle is less than one degree its dimension should begin with 0°, for example, 0° 30′.

Leader lines

Notes are often used on drawings together with a **leader line** which shows where the note applies. Leader lines are thin continuous lines which end in an arrowhead or a dot, as shown in Figure 3.4(a) and (b). Leader lines with arrowheads must touch and stop on a line, as in Figure 3.4(a); those with dots should have the dot within the outline, as in Figure 3.4(b).

(a)

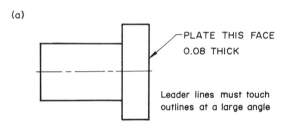

Leader line with arrowhead

(b)

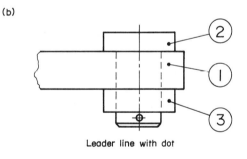

Leader line with dot

(c)

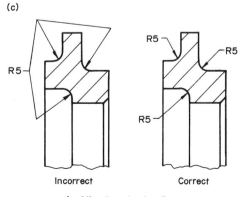

Avoiding long leader lines

Figure 3.4 *Leader lines*

Figure 3.4(b) also shows 'balloons' containing item references (1, 2, 3, etc.). These should be arranged in vertical or horizontal groups.

To avoid any confusion when reading the drawing, leader lines must not touch outlines at an acute angle, neither should they be parallel to nearby dimension or projection lines. Long leader lines should be avoided. Figure 3.4(c) shows how this may be done.

Arrangement of dimensions

Dimensions should be placed near the middle of the dimension line, and above and clear of it as illustrated in Figure 3.5. They should be placed so that they are not crossed by any other line of the drawing.

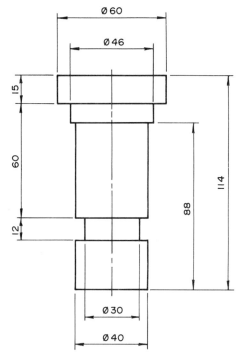

Figure 3.5   *Dimensions and dimension lines*

For easy reading, dimensions should be placed so that they can be read from the bottom or right-hand side of the drawing. Figure 3.6 shows the application of this rule to linear dimensions, and Figure 3.7 its application to angular dimensions. Wherever possible, dimensions should be placed outside the hatched zones as in Figure 3.6. Figure 3.7 shows that the dimension line for an angle is a circular arc with its centre on the point of the angle.

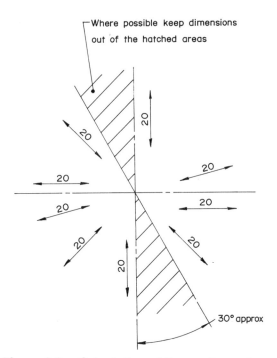

Figure 3.6   *Orientation of linear dimensions*

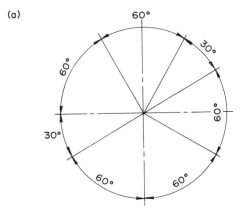

(a)

Dimensions arranged to read from bottom or right-hand side

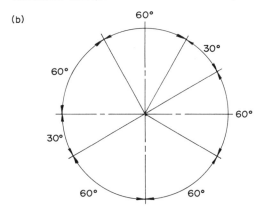

(b)

Dimensions arranged horizontally

Figure 3.7   *Orientation of angular dimensions*

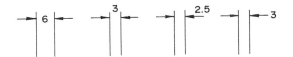

Figure 3.8 *Dimensioning small features*

Where small features have to be dimensioned and there is limited space in which to place the dimension, any of the methods in Figure 3.8 may be used.

Auxiliary dimensions

In Figure 3.5 the overall length, 114, of the shaft is given, together with three intermediate length dimensions, 12, 60 and 15. The remaining length dimension, that of the 40 diameter, is redundant and is therefore omitted. If it were given, then the overall length would be shown in two different ways, which could cause confusion.

However, redundant dimensions will sometimes provide useful information. In this case they are given as **auxiliary dimensions**. Where all the intermediate dimensions are necessary, the overall dimension may be shown as an auxiliary dimension by enclosing it in brackets. Figure 3.9 is an illustration of this. The overall length of a part is sometimes needed to decide how much material to order, or to cut stock into approximate lengths. An auxiliary dimension is not used in the machining or inspection of a part.

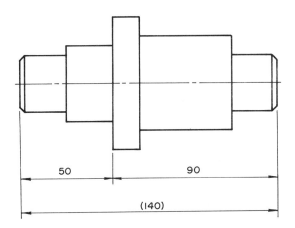

The overall length has been given as an auxiliary dimension

Figure 3.9 *Auxiliary dimension*

Dimensioning methods

Parallel dimensioning. With parallel dimensioning, shown in Figure 3.10, a number of dimensions originate from a **datum feature** which is generally a face, an edge or an axis. In the figure the left-hand end of the part is the datum feature for positioning the three hole axes. See also Figure 3.30(b) and Figure 3.31.

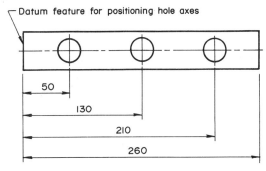

Figure 3.10 *Parallel dimensioning*

Superimposed running dimensioning. This is a simplified form of parallel dimensioning. It is useful when there is insufficient space for all the parallel dimension lines. The datum feature is identified by the **common origin indicator**, as shown in Figure 3.11. This illustrates two positions for the dimensions. Figure 3.11(a) is the usual form, but the positioning in Figure 3.11(b) is useful where there are only small differences in the dimensions leaving insufficient space to insert the values.

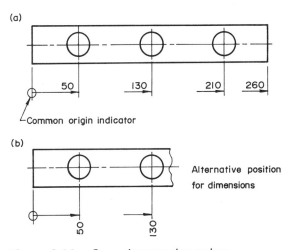

Figure 3.11 *Superimposed running dimensioning*

Chain dimensioning. An example of this method of dimensioning is given in Figure 3.12. Chains of dimensions should be used only where the possible accumulation of tolerances over the length of the part does not endanger the correct functioning of the part. See also Figure 3.30(a).

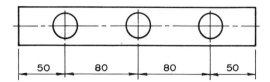

Figure 3.12 *Chain dimensioning*

Combined parallel and chain dimensioning.
This dimensioning method, shown in Figure 3.13, is
often used. In the figure the left-hand end of the part
is the datum feature for positioning the axes of the
larger holes. These in turn are the datum features for
the axes of the smaller holes.

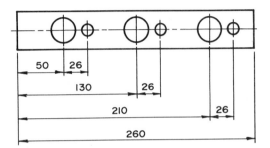

Figure 3.13 *Combined parallel and chain
dimensioning*

Dimensioning by co-ordinates. In Figure 3.14
superimposed running dimensioning has been used
in two directions at right angles. The common origin
in this case is the bottom left-hand corner of the part,
but any suitable datum feature may be used, such as
the axis of one of the holes.

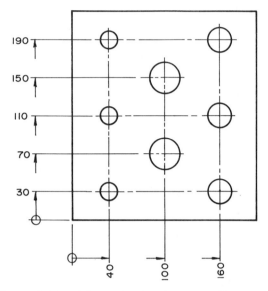

Figure 3.14 *Dimensioning by co-ordinates*

Self-assessment questions

3.1 Projection and dimension lines are thick.   True or false?

3.2 What is the usual angle between the projection and dimension lines of a dimension?

3.3 What are the recommended maximum and minimum lengths for arrowheads?

3.4 Relative to the outline of the view, where should leader lines ending in a dot finish?

3.5 Dimensions are always arranged to read from the bottom of the drawing.   True or false?

3.6 How is an auxiliary dimension indicated on a drawing?

3.7 What disadvantage is there is using chain dimensioning?

Symbols and abbreviations

Symbols and abbreviations are used on drawings to
save space and time. Provided those used are
commonly understood they also simplify the
dimensioning and at the same time give precise and
clear descriptions.

A selection of commonly used and understood
symbols and abbreviations is given in Table 3.1.
Others should not be used; instead, the intended
meaning should be stated in words.

Abbreviations are the same in the singular and
plural, that is, for example, 'drawing' and 'drawings'
are both shortened to DRG. An abbreviation does
not end with a full stop, except where it makes a
word itself, as with NO. for number.

Table 3.1 Commonly accepted symbols and abbreviations

Term	Abbreviation or symbol
Across flats	AF
Assembly	ASSY
Centres	CRS
Centre line	
on a view	℄
in a note	CL
Chamfer, chamfered (in a note)	CHAM
Cheese head	CH HD
Counterbore	CBORE
Countersink	CSK
Diameter (in a note)	DIA
Diameter (preceding a dimension)	Ø
Drawing	DRG
Equally spaced	EQUI SP
Hexagon	HEX
Hydraulic	HYD
Insulated, insulation	INSUL
Left-hand	LH
Long	LG
Material	MATL
Maximum	MAX
Minimum	MIN
Number	NO.
Pitch circle diameter	PCD
Radius (in a note)	RAD
Radius (preceding a dimension)	R
Required	REQD
Right-hand	RH
Sheet	SH
Specification	SPEC
Spherical diameter (only preceding a dimension)	SØ
Spherical radius (only preceding a dimension)	SR
Spot face	SFACE
Square (in a note)	SQ
Square (preceding a dimension)	□
Taper, on diameter or width	▷
	(orientated to direction of taper)
Thread	THD
Undercut	UCUT

Methods of dimensioning common features

Circles and diameters. Complete circles must always be dimensioned by their diameters using one of the methods shown in Figure 3.15. The dimension is preceded by the symbol Ø, meaning diameter. Circles must be shown with two centrelines.

Figure 3.16 shows several ways in which diameters may be dimensioned. The dimension should be placed on the view which ensures the maximum clarity, as in Figure 3.16(a). Here the dimensions of the diameters are placed on the lengthwise view rather than on the other view, which has a number of concentric circles. In Figure 3.16(b) some of the diameter dimensions are related to the features by leader lines. This avoids having a number of projection lines on the view. The methods in Figure 3.16(c) are useful where space is restricted.

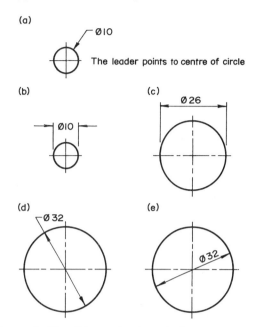

Figure 3.15 *Dimensioning circles*

The diameter of a spherical surface should be dimensioned as in Figure 3.17.

Radii. Radii are dimensioned using a dimension line which passes through, or is in line with, the centre of the arc. The dimension line carries one arrowhead only and this touches the arc. The symbol R, meaning radius, precedes the dimension. These points are illustrated in Figure 3.18, where Figure 3.18(a) shows the dimensioning for arcs whose centres are located, and Figure 3.18(b) the dimensioning for arcs which do not need to have their centres located.

Where the size of a radius is fixed by another dimension, as in Figure 3.18(c), this is indicated by the symbol R without a dimension. In the example in Figure 3.18(c), the width of the features is given as 60, so the radius must be 30.

Figure 3.19 shows how spherical radii are dimensioned.

Chamfers. Chamfers at 45° should be dimensioned using one of the methods in Figure 3.20(a). For chamfers with angles other than 45° the dimensions should be shown as in Figure 3.20(b).

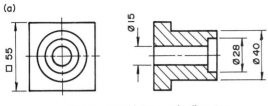
(a)

Closely-spaced concentric diameters

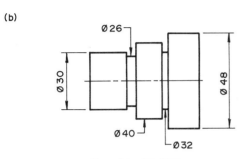
(b)

Use of leader lines

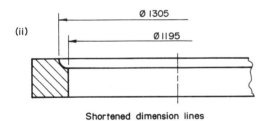
(c)

(i)

(ii)

Shortened dimension lines

Figure 3.16 *Dimensioning diameters*

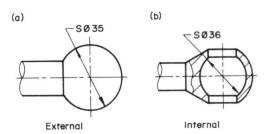
(a)   (b)

External   Internal

Figure 3.17 *Dimensioning spherical diameters*

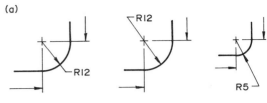
(a)

Arcs with centres located

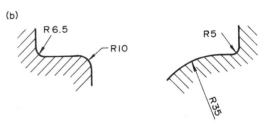
(b)

Arcs with centres not located

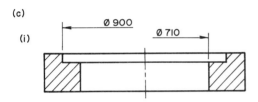
(c)

Radii established by other dimensions

Figure 3.18 *Dimensioning radii of arcs*

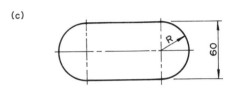
(a)   (b)

Figure 3.19 *Dimensioning spherical radii*

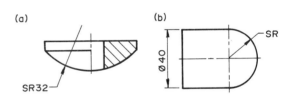
(a)

45° chamfers

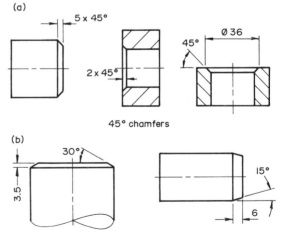
(b)

Chamfers at angles other than 45°

Figure 3.20 *Dimensioning chamfers*

Size of holes. Figure 3.21 shows typical methods of dimensioning holes. In general, production methods such as drill, ream, punch, etc., are not specified. When the depth of holes is given in a note, as in Figure 3.21(b), the depth does not include the conical bottom of the hole, which is produced by the end of the drill.

(a)

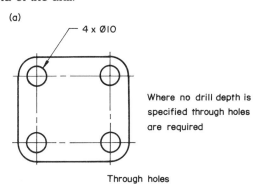

Where no drill depth is specified through holes are required

Through holes

(b)

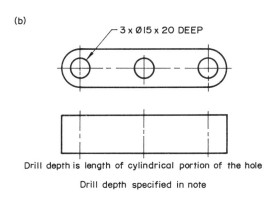

Drill depth is length of cylindrical portion of the hole

Drill depth specified in note

(c)

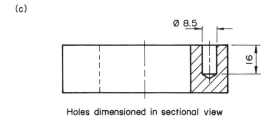

Holes dimensioned in sectional view

Figure 3.21 *Dimensioning holes*

Hole positioning. Holes and other features should be positioned by spacing them on circles, as shown in Figure 3.22, or by giving them rectangular co-ordinates or centre distances, as in Figure 3.23. If the holes are positioned by co-ordinates, although they do in fact lie on an arc or circle as in Figure 3.23, the pitch circle radius or pitch circle diameter should be given as an auxiliary dimension.

Countersinks. When parts are fixed together using countersunk screws, the holes for the screws have a conical end, called a **countersink**, into which

(a)

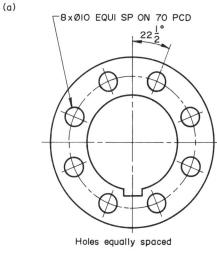

Holes equally spaced

(b)

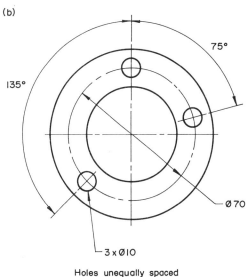

Holes unequally spaced

Figure 3.22 *Dimensioning hole positions on a circle by angular spacing*

the screw heads fit. These may be dimensioned by one of the methods in Figure 3.24.

Counterbores. When a hole is **counterbored**, one end is enlarged as shown in Figure 3.25. The figure illustrates ways in which counterbores are dimensioned. The length of the hole after counterboring should be given, as in Figure 3.25(b), if it is more important than the depth of the counterbore.

Spotfaces. Where it is necessary to clean up a circular area around the end of a hole, called a **spotface**, the dimensioning should use one of the methods in Figure 3.26. Spotfaces are often used on castings to provide a flat surface on which a bolt head may seat. No depth is specified for a spotface.

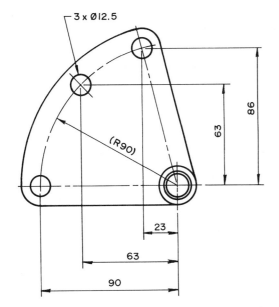

Pitch circle radius given as auxiliary dimension

Figure 3.23 *Dimensioning hole positions by co-ordinates*

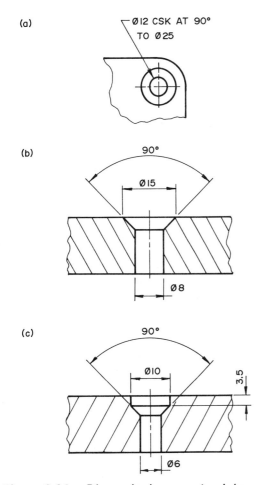

Figure 3.24 *Dimensioning countersinks*

(a)

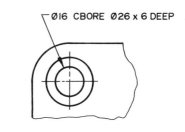

(b)

(c)

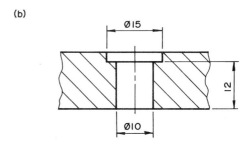

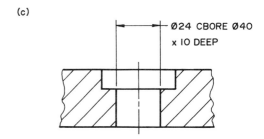

Figure 3.25 *Dimensioning counterbores*

(a)

(b)

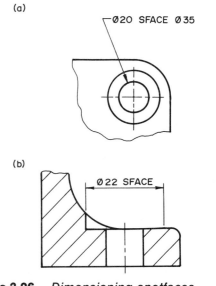

Figure 3.26 *Dimensioning spotfaces*

58

The depth is required to be the minimum necessary to provide a completely machined surface of the specified diameter.

Tapered features. Tapered features may be dimensioned by different methods as shown in Figures 3.27 and 3.28, the method used depending on the function of the part. Figure 3.28 illustrates the use of the international symbol for taper.

The dimensioning of screw threads is shown in Figure 4.4, and that for keyways in Figure 5.8.

(c)

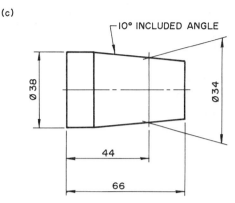

Figure 3.27 *Dimensioning tapered features*

(a)

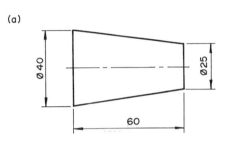

(b)

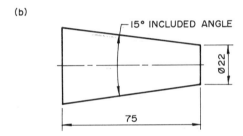

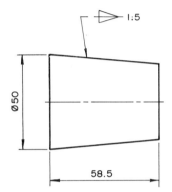

The taper symbol is recognised internationally and points in the direction of the taper

Figure 3.28 *Taper symbol*

Self-assessment questions

3.8   An assembly requires six square-headed pins, 16 diameter and 25 long. Using appropriate symbols and abbreviations, write a note to appear on a drawing which will describe these parts.

3.9   Write down the terms which the following abbreviations represent.

AF;   HEX HD;   INSUL;   RH;   HYD;   LG;   SH;   THD;   SR;   CRS.

3.10   Circles may be dimensioned using their diameters or their radii.   True or false?

3.11   (a) The end of a shaft 35 diameter is to carry a 45° chamfer which is 3 long. Make a dimensioned sketch of the chamfer.
(b) The angle of the chamfer is changed to 35° measured to the end of the shaft. Make a second dimensioned sketch of the new chamfer.

3.12   A thin gasket has an outside diameter of 100 and a bore of 60. There are six holes in the gasket of 12 diameter, equally spaced on a pitch circle diameter of 80. Make a dimensioned drawing of the gasket.

3.13   A part 35 thick and 60 diameter has a 20 diameter central hole through it, one end of which has a 90° countersink. The diameter at the end of the countersink is 38. Make a dimensioned drawing of the part, in section through the hole.

59

Tolerances

No manufactured item is ever exactly the same as any similar item because of the inaccuracies in the machine or manufacturing process, tool wear, temperature changes and so on. So the designer has to decide the maximum and minimum sizes for dimensions which will enable the part to function satisfactorily. The difference between the maximum and minimum sizes is the amount within which the dimension may vary. It is called the **tolerance**. Figure 3.29 shows ways in which tolerances may be expressed.

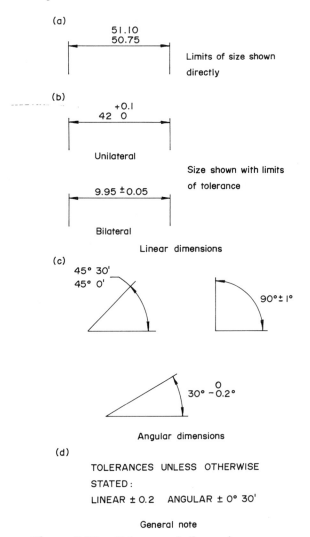

Figure 3.29 *Toleranced dimensions*

Linear dimensions. In Figure 3.29(a) the maximum and minimum sizes, or **limits of size**, are shown directly. The larger limit of size is placed above the smaller and both are given to the same number of decimal places.

Figure 3.29(b) shows a second method of tolerancing linear dimensions. A size is stated with **limits of tolerance** above and below that size. Both limits of tolerance have the same number of decimal places, except when one limit is nil, when it is written '0' without a plus or minus sign. It is not necessary to give the dimension to the same number of decimal places as the limits of tolerance. As with the first method, the limit of tolerance which gives the larger limit of size is placed above that which gives the smaller.

Figure 3.29(b) also illustrates the difference between **unilateral** and **bilateral** tolerances. A unilateral tolerance lies entirely on one side of the dimension while bilateral tolerances lie partly on each side of the dimension.

Angular dimensions. These are toleranced using the same methods as for linear dimensions. Examples are shown in Figure 3.29(c).

General tolerances. Except for auxiliary dimen-

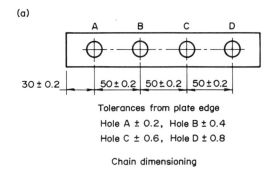

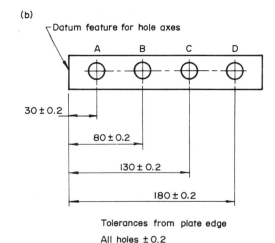

Figure 3.30 *Cumulative effects of tolerances*

sions, a dimension without a tolerance is incomplete. The tolerance may not appear which the dimension however. If some dimensions of a part do not need to be controlled with special accuracy the tolerance may be shown as a general note. Figure 3.29(d) is an example.

Single limits of size.   Sometimes only one limit of size for a dimension need be specified. Examples are the minimum length of thread in a tapped hole, or the maximum size of a radius permitted at a corner. In such cases the abbreviation MIN or MAX should follow the dimension.

Cumulative effects of tolerances.   Where chain dimensioning of toleranced centre distances is used, as in Figure 3.30(a), the tolerances may be cumulative, that is they add up, as shown, This accumulation of tolerance can be reduced by using parallel dimensioning, as shown in Figure 3.30(b).

Size of tolerances.   Sometimes beginners in draughting think that specifying small tolerances for a part will lead to increased accuracy in manufacture. However, the largest tolerances which will provide satisfactory functioning should be used. This is because the cost of production of a part increases as the size of the tolerances decreases. Small tolerances require higher skill from the machine operator, more frequent resetting of the machine because of tool wear, and more time to complete the manufacture.

Self-assessment questions

3.14   All dimensions on a drawing must have tolerances.   True or false?

3.15   The maximum and minimum sizes for a hole are 19.9 diameter and 20.1 diameter. Which of these should be placed above the other when the hole is dimensioned on the drawing?

3.16   Two dimensions on a drawing are shown as 35.5 ± 0.5 and 25 MIN. Which has a unilateral tolerance and which a bilateral?

3.17   An angular feature on a drawing is to have a maximum angle of 61° and a minimum of 60°. How could this requirement be shown with (a) a unilateral tolerance, and (b) a bilateral tolerance?

3.18   The functioning of a part will always be improved if the tolerances on the dimensions are reduced.   True or false?

The selection of dimensions

General

All dimensions which are needed to define the part completely must be shown on the drawing. Each dimension must appear only once, or confusion about which is correct will be caused. A dimension should not have to be calculated from other dimensions, and the drawing should not have to be scaled. Do not show any dimensions which are not needed to define the part.

All dimensions for a particular feature should preferably be shown on a single view rather than on several. The view used should be that which shows the typical shape of the feature. For example, in Figures 3.22 and 3.23, the holes and their positions are given on the view which shows the holes as circles. Dimensions should not be placed on a view where the feature is shown in hidden detail. If it is essential to dimension such a view it should be drawn in section.

Size and location dimensions

In general, dimensions fall into two main groups: those which show the sizes of the feature; and those which position the feature relative to other features. In Figure 3.23 for example, the note '3 × Ø12.5' gives the size of the three holes, and the vertical and horizontal dimensions locate the holes from the axis of the lower right-hand bore.

When dimensioning a drawing, the sizes of the features are first determined and inserted on the

(a)

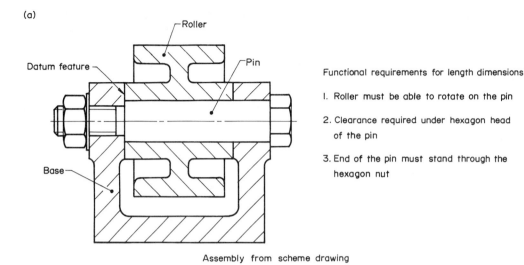

Functional requirements for length dimensions

1. Roller must be able to rotate on the pin

2. Clearance required under hexagon head of the pin

3. End of the pin must stand through the hexagon nut

Assembly from scheme drawing

(b)

f^1 f^2 (bigger than f^5)

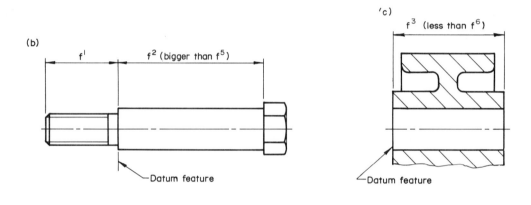

Datum feature

'c)

f^3 (less than f^6)

Datum feature

(d)

f^4 f^5

f^6

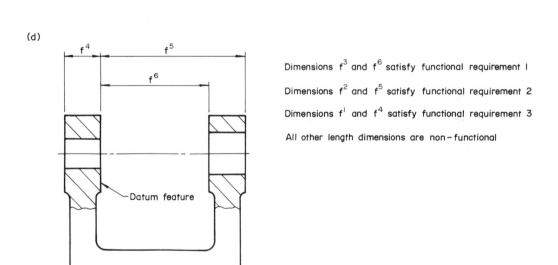

Datum feature

Dimensions f^3 and f^6 satisfy functional requirement 1

Dimensions f^2 and f^5 satisfy functional requirement 2

Dimensions f^1 and f^4 satisfy functional requirement 3

All other length dimensions are non-functional

Functional length dimensions for parts

Figure 3.31 *Functional dimensions*

drawing. Methods for dimensioning common features are given earlier in this chapter. Then location dimensions are added to position the features relative to each other.

Cylinders, tapers, and other symmetrical features are located by their axes, not their surfaces. Castings and forgings are only partially machined. Their location dimensions should be taken from those machined surfaces, or the axes of machined bores. Figure 6.11(b) shows the symbol to be used to indicate that a surface is to be machined.

Functional dimensions

As stated above, for the drawing of a part to be dimensioned satisfactorily the dimensions must define the part completely and unambiguously, without the need for the workshop to make calculations or to refer doubtful points to the designer for clarification.

Equally however, satisfactory dimensioning requires that dimensions which are essential to the correct functioning of the part, called **functional dimensions**, are stated directly on the drawing. The remaining non-functional dimensions can be chosen to aid the production or inspection of the part.

Figure 3.31 is an example of the selection of functional length dimensions. Figure 3.31(a) shows a roller assembly as it would appear on a scheme or design drawing. From this the datum feature for the functional length dimensions can be deduced. The shoulder on the threaded end of the pin is pulled up against the inside face of the left-hand boss on the base when the hexagon nut is tightened. To ensure that the shoulder is pulled up to the boss face there must be some small clearance under the hexagon head of the pin. Furthermore, the roller must be able to rotate on the pin.

The inside face of the left-hand boss on the base is therefore the datum feature for this part. The shoulder on the pin is the datum feature for the pin. The datum feature for the roller is the face of the left-hand boss.

Figure 3.31(b), (c) and (d) shows the required functional dimensions on the parts. All of them originate from the datum features and are stated directly. Dimension f^2 on the pin must be bigger than dimension f^5 on the base, and dimension f^3 on the roller must be less than dimension f^6 on the base. This can be arranged by choosing suitable tolerances for the dimensions.

Self-assessment questions

3.19 Should important dimensions be repeated several times to make the drawing easy to read?

3.20 Generally, is it best to show all the dimensions for a feature on one view?

3.21 Four equally spaced holes 10 diameter and 15 deep are to be drilled in a component on a 60 diameter pitch circle. The centre of the pitch circle is to be 80 from a face on the part. Which of these figures are size dimensions and which location dimensions?

Exercise 3.1

Make correctly dimensioned drawings from the following instructions.

a) Seven holes 10 diameter are spaced 20 apart along a straight line. The axis of the left-hand end hole is the common origin for the spacing.

b) Six 12 diameter holes are equally spaced on a 60 diameter circle. The axis of the top hole is positioned 15° to the right of the vertical centreline of the 60 diameter circle.

c) An 8 diameter hole is drilled through a plate 20 thick. The top of the hole is enlarged to 12 diameter for a depth of 4. At the other end of the 8 diameter hole a 90° countersink is machined to a diameter of 10. A sectional view through the hole is required.

d) A tapered bore is machined in a part which is 80 thick. The large end of the hole is 30 diameter and the taper is 1 : 4. A sectional view through the bore is required.

Exercise 3.2

A conical spacer is 70 long with a large-end diameter of 40 and a small-end diameter of 30. Axially through the centre a 10 diameter hole is drilled. At the large end this hole is counterbored 18 diameter by 15 deep. Draw a dimensioned view in section through the centre line.

Exercise 3.3

A retaining plate is square with sides 140. Its thickness is 29.5/30.0. The corners of the square have 20 radii and at the centres of these radii four 20 diameter holes are drilled through the plate. These holes are spotfaced to 30 diameter at one end. In the centre of the plate is a 59.95/60.00 diameter hole with a 45° chamfer 3 deep at each end. Draw two dimensioned views of the plate, one in section through the central hole. Except where stated, linear dimensions have a tolerance of ±0.2 and angular dimensions ±0° 30'.

Exercise 3.4

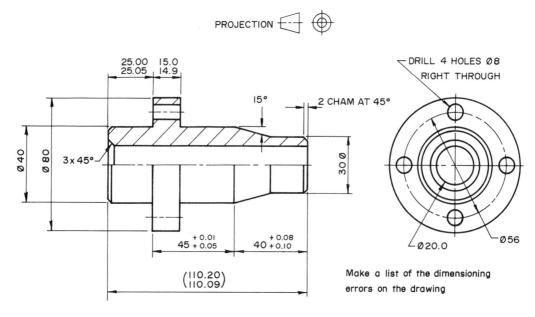

Make a list of the dimensioning errors on the drawing.

Exercise 3.5

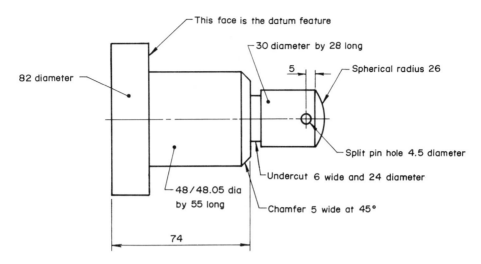

Copy the drawing and dimension it correctly.

Exercise 3.6

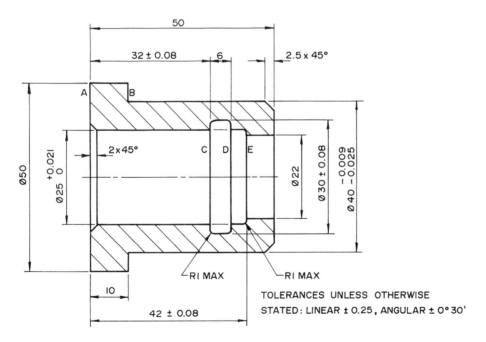

(a) Identify the datum face.

(b) Calculate maximum and minimum distances between faces A and D, B and C, and C and E.

(c) The Ø40 feature is to fit in a bore with a maximum clearance of 0.08 and a minimum clearance of 0.005. Find the maximum and minimum sizes of the hole.

Exercise 3.7

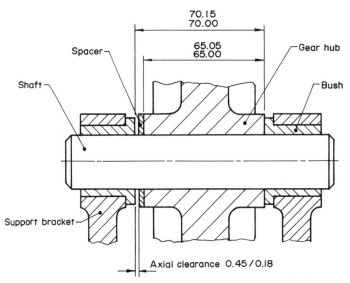

(a) Find the limits of size for the thickness of the spacer to provide the required axial clearance of 0.45/0.18.

(b) The design is changed and one of the spacers is used each side of the gear hub. The axial clearance is now to be 0.38/0.12. Find the new limits of size for the length of the gear hub.

Exercise 3.8

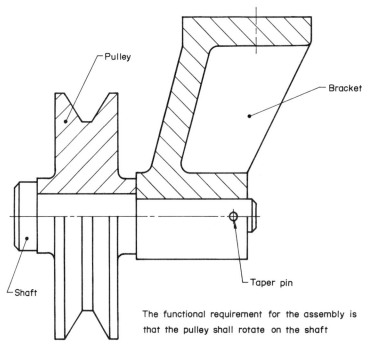

The functional requirement for the assembly is that the pulley shall rotate on the shaft

Determine the datum feature for the assembly and sketch each part showing the functional dimensions.

Post test 3

1 (a) For neatness, leader lines are drawn either horizontally or vertically.   True or false?
   (b) SR preceding a dimension means 'spherical radius'.   True or false?
   (c) An angle of fifty-five minutes thirty seconds should be dimensioned as 55' 30''.   True or false?
   (d) The diameter and depth of a spotface are needed to specify it completely.   True or false?
   (e) The dimension line for a radius has only one arrowhead.   True or false?
   (f) The maximum and minimum permitted values for a dimension are the limits of size.   True or false?
   (g) Non-functonal dimensions do not need tolerances.   True or false?

2 (a) Calculate the maximum and minimum distances from the datum feature to slots B and D.
   (b) Redraw the part with slot A as the datum feature. Change the dimensioning so that the tolerance between any pair of slots is ±0.1.

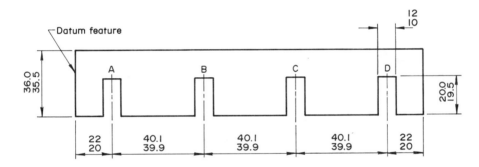

3 Make sketches of the parts of the assembly and identify the functional dimensions necessary to achieve the specified minimum clearance.

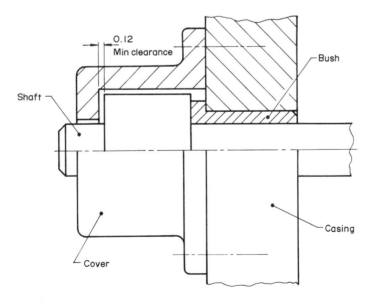

Screw threads, screw fasteners and locking devices

Principal learning objective

The student uses the conventions in British Standards for representing screw threads, screw fasteners and locking devices.

Conventional representation of screw threads

A screw thread is a helix. To draw a true representation of it would take a considerable time, so for all normal purposes a conventional representation is used, whatever the type of screw thread being drawn.

External threads

Figure 4.1(a) shows how external threads are represented on an outside view. The major diameter is shown by a pair of thick lines, and the minor diameter by a pair of thin lines. The limit of useful

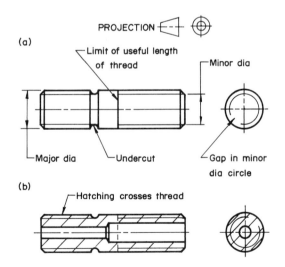

Note use of thick and thin lines

Figure 4.1 *Conventional representation of external screw threads*

length of thread is drawn as a thick line. This line shows the end of the full thread, that is, the point at which the thread ceases to be fully formed. Beyond this point is the thread run-out, consisting of incompletely formed threads produced by the lead-in chamfer on the die. On the circular view the major diameter appears as a complete thick circle and the minor diameter as a thin circle with a gap.

When external threads are drawn in section, as in Figure 4.1(b), the minor diameter is again represented by a pair of thin lines. The hatching crosses those lines and ends at the major diameter.

Internal threads

For an outside view of an internal thread all the lines are hidden and therefore are drawn dashed. An internal thread may be either a **blind tapping**, as in Figure 4.2(a), or a **through tapping** as in Figure 4.2(b). The minor diameter drilling is drawn in thick lines, as is the line representing the end of the full thread in a blind tapping. The included angle of the tapping drill point is 120°. The hatching crosses the thread, in the same way as for an external thread in section. The circular view shows the minor diameter as a complete thick circle and the major diameter as a thin circle with a gap.

Conventions for assembled threads

The general rule is that external threads are shown covering internal threads, whether the external thread is shown in section or not. In Figure 4.3(a) the part with the external thread is unsectioned, so its

PROJECTION ⊏⊐⊦ ⊕

(a)

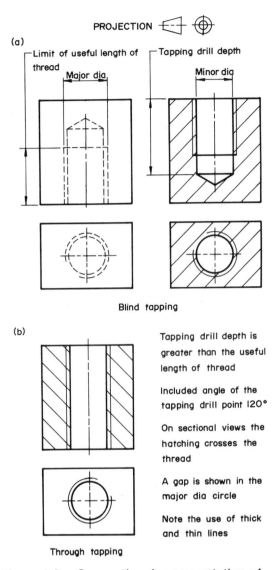

Limit of useful length of thread

Major dia.

Tapping drill depth

Minor dia.

Blind tapping

(b)

Through tapping

Tapping drill depth is greater than the useful length of thread

Included angle of the tapping drill point 120°

On sectional views the hatching crosses the thread

A gap is shown in the major dia circle

Note the use of thick and thin lines

Figure 4.2 *Conventional representation of internal screw threads*

(a)

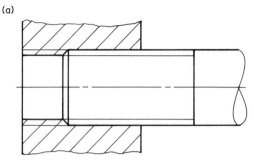

Unsectioned external thread

(b)

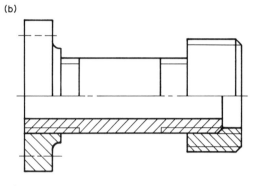

Sectioned external thread

External threads drawn covering internal threads

Figure 4.3 *Conventions for assembled screw threads*

major diameter is shown as thick lines and its minor diameter as thin lines. The reverse applies to the sectioned internal thread, and the hatching crosses it.

When the external thread is sectioned, as in Figure 4.3(b), the minor diameter is shown as thin lines and the hatching crosses them. Note that the external threads cover the internal threads, as stated above.

Dimensioning of screw threads

Since threads are shown conventionally on engineering drawings they must be designated completely by dimensions or a note. The most commonly used thread is the ISO metric thread and this is designated as:

M16 × 1.5 – 6H for an internal thread
M6 × 0.75 – 6g for an external thread

M is the thread system symbol for ISO metric, ISO standing for International Organization for Standardization. 16 and 6 are the nominal diameters (major diameters) in millimetres.

1.5 and 0.75 are the thread pitches in millimetres. 6H and 6g are the thread tolerance class symbols. The only difference between the designations of internal and external threads is in the tolerance class symbol.

In many countries which use the ISO metric thread the convention is to omit the pitch from the designation. If no pitch is shown a thread from the coarse-pitch series is implied. Thus a coarse thread M6 × 1 – 6H may be shown as M6 – 6H. Fine-pitch threads are used mainly in special applications such as for thin-walled components and fine-adjustment machine tools.

The internal and external thread tolerances provide three classes of fit: medium, close and free. The medium fit (6H/6g) is suitable for most general engineering purposes. The close fit (5H/4h) is used when close accuracy of thread form and pitch is needed, but requires high quality equipment to produce the threads and particularly thorough inspection. The free fit (7H/8g) is for use in applications needing quick and easy assembly, even if the threads are dirty or slightly damaged.

Figure 4.4 shows examples of screw thread dimensioning. Figure 4.4(a) and (b) illustrate the treatment when the dimension note is applied to the circular view of the thread. Figure 4.4(c), (d) and (e) show methods for dimensioning internal threads, and Figure 4.4(f) and (g) show how the length of the thread may be controlled where necessary. Figure 4.4(f) also shows the conventional representation for the thread run-out. The run-out is shown as thin lines at 30° to the major diameter, but only if the run-out is necessary for the part to function correctly. See Figure 4.6(c) for an example.

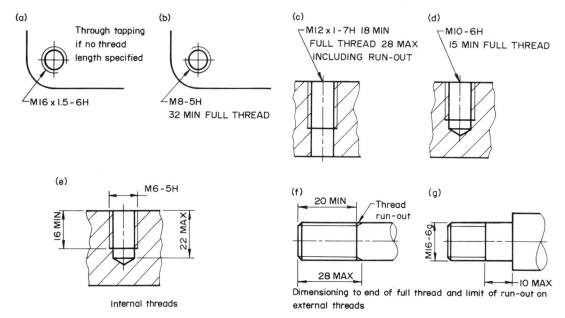

Figure 4.4 *Dimensioning of screw threads*

Self-assessment questions

4.1   When external threads are drawn the minor diameter is always represented with thin lines. True or false?

4.2   For a blind tapping the useful length of thread and the tapping drill depth are always the same. True or false?

4.3   The included angle of the tapping drill point is always shown as 90°.   True or false?

4.4   When assembled threads are drawn, external threads are shown covering internal threads. True or false?

4.5   A screw thread on a drawing is dimensioned as M20 × 1.5 – 5H. Write down what each of these terms means, state whether the thread is internal or external and whether it is a coarse- or fine-pitch thread.

Exercise 4.1

PROJECTION

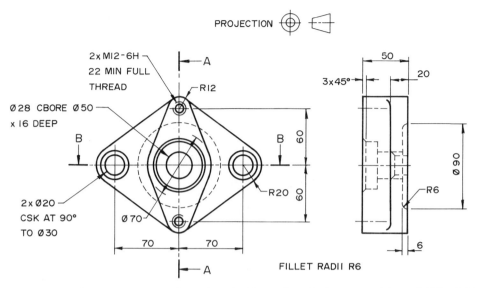

Draw the left-hand view and project half sectional views on AA and BB. Add six main dimensions.

Exercise 4.2

PROJECTION

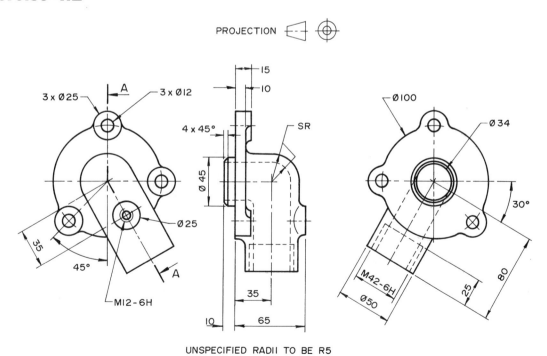

UNSPECIFIED RADII TO BE R5

Draw the left-hand view and project a sectional view on A–A.

Exercise 4.3

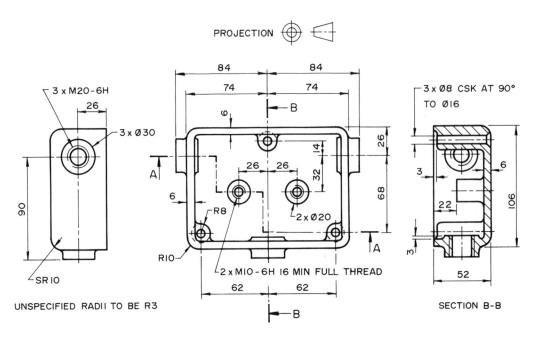

PROJECTION

Draw the centre view and project a sectional view on A–A.

Exercise 4.4

PROJECTION

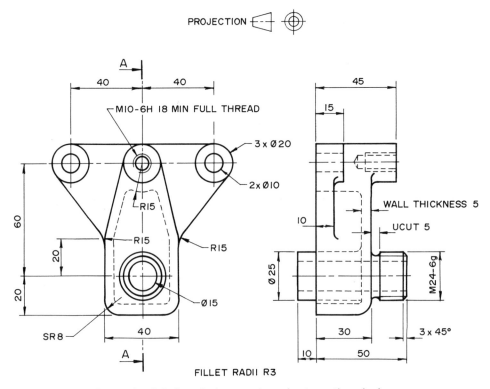

Draw the left-hand view and project section A–A.

Common screw fasteners

Hexagonal bolts and nuts

The commonest form for nuts, and bolt and screw heads is hexagonal with a 30° chamfer to remove the sharp corners at the top of the bolt head and usually at each end of the nut. In the hexagonal view of bolt heads and nuts this chamfer appears as a circle with a diameter of 1.5 times the major diameter, d, of the thread. Tangential to this circle the hexagon is drawn using a 30°–60° set square. From this view the other view of the fastener can be projected, as shown in Figure 4.5(a) and (b). These illustrate the terms associated with bolts and screws and with nuts, and show that the thickness of the bolt head is drawn as $0.7d$ and that of a nut as $0.8d$.

The curves on the flats of the hexagon can be drawn with compasses or a radius template. They should approximate to those shown in Figure 4.5(a) and (b), which were drawn using a nut and bolt template. The only difference between a bolt and a screw is that a screw is threaded for virtually its entire length. The first thin thread on each is removed by a 45° chamfer extending to the minor diameter.

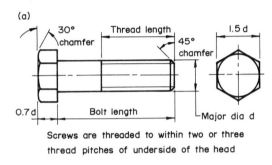

Screws are threaded to within two or three thread pitches of underside of the head

Hexagon bolts and screws

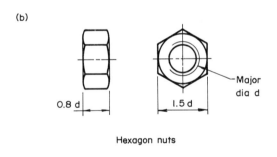

Hexagon nuts

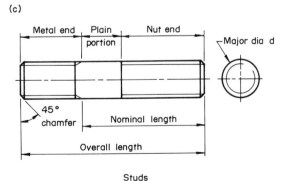

Studs

Figure 4.5 *Common screw fasteners*

Studs

Figure 4.5(c) shows a stud and the terms associated with this type of screw fastener. A stud is a length of circular section bar, threaded at each end with a plain portion between. One end, called the **metal end** or **fast end**, is screwed into one detail of the assembly. The other end, called the **nut end**, passes through a plain hole in the other part of the assembly and carries the nut which locks the assembly together.

The metal-end threads are made either oversize or to tolerance class 4h (close fit), so that they are tight in the tapping. The nut-end threads are made to tolerance class 6g (medium fit). The use of different tolerance classes ensures that when the nut is removed the stud is not unscrewed as well.

Note that the thread run-out at the metal end is shown because it helps to secure the stud by being tight in the first threads of the tapping.

The length of the metal-end thread, which includes the run-out threads, is d or $1.5d$, where d is the major diameter of the thread. The shorter length is used for steel, the longer for soft or brittle materials such as aluminium alloy or cast iron.

Studs are used where there is insufficient space on one side of the assembly for bolt heads or nuts. They are also used to secure parts such as inspection covers which have to be removed frequently. The use of a nut and bolt in such an application would require two spanners to make the joint. If a set bolt were used, see Figure 4.6(b), the frequent removal and replacement would quickly wear the thread in the tapping, particularly if the material were cast iron or a light alloy.

Self-assessment questions

4.6 What is the drawing size for the diameter of the chamfer circle on a hexagon head bolt?

4.7 What is the drawing size for the thickness of a hexagon nut?

4.8 What is the difference between a bolt and a screw?

4.9 A stud is threaded for its entire length. True or false?

4.10 Give two reasons for using studs and nuts to make a joint rather than bolts and nuts.

Nut and bolt assembly

Figure 4.6(a) shows a sectional view through such an assembly. Normally the parts have clearance holes for the bolt to ensure easy assembly. The bolt should be long enough to project right through the nut, so that all the nut threads are engaged and all are sharing the tensile load in the bolt. Note that the nut and bolt are not sectioned although the cutting plane passes through them.

If the details being joined are of a soft material a plain washer may be placed under the nut, and sometimes under the bolt head as well, to prevent damage when the nut is tightened.

Set bolt assembly

A set bolt is sometimes called a **tap bolt**. It secures two parts together by being screwed into a tapping in the inner, as shown in Figure 4.6(b), the outer part having a clearance hole for the bolt.

To clamp the parts together the underside of the bolt head must bear on the outer part. To ensure this, the end of the bolt must not reach the bottom of the

(b)

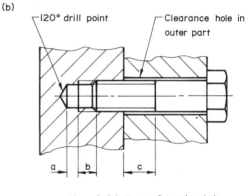

a – unthreaded bottom of tapping hole
b – thread in tapping hole below bolt
c – bolt thread above joint line

Set bolt assembly

(c)

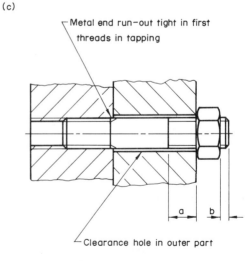

a and b — Nut end thread each side of nut

Nut and stud assembly

(a)

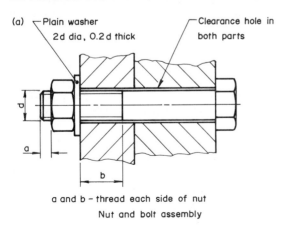

a and b – thread each side of nut

Nut and bolt assembly

Figure 4.6 *Screw fastener assemblies*

thread in the tapping, and some bolt threads must be above the joint line between the parts.

Nut and stud assembly

A sectional view through an example of this type of assembly is shown in Figure 4.6(c). The stud is screwed into the tapping until the run-out threads at the metal end jam in the first thread in the hole. The outer part has a clearance hole, as with a set bolt, and the nominal length of the stud should be sufficient for it to stand through the nut.

Exercise 4.5

Two pieces of material each 50 thick are to be secured together by an M16 hexagon bolt and nut. Draw a sectional view of the assembly similar to Figure 4.6(a). Show a plain washer under the nut.

Exercise 4.6

Draw a sectional view of a set bolt assembly as shown in Figure 4.6(b), with the following dimensions: Tapped part 50 thick, outer part 32 thick. Tapping M16, 28 minimum length of full thread. Other dimensions to be settled by the student.

Exercise 4.7

Two pieces of light alloy, each 20 thick, are to be joined by an M6 stud and hexagon nut in a blind tapped hole. The nut is to be completely housed in a counterbore. Using suitable dimensions draw twice full size a sectional view of the assembly similar to that in Figure 4.6(c). Show a plain washer under the nut.

Note that the diameter of the counterbore must be big enough to accommodate a tubular spanner to tighten the nut.

Other head forms for screws

Figure 4.7 shows some other common head forms and approximate drawing proportions for screw fasteners. They are all known as screws, whether they are threaded up to the head or not.

Screws with hexagon socket heads are tightened by a special key or wrench made from hexagonal bar bent through 90° and called an **Allen key**. Cap screws with hexagon socket heads are often called Allen screws and are used extensively on machine tools and jigs and fixtures.

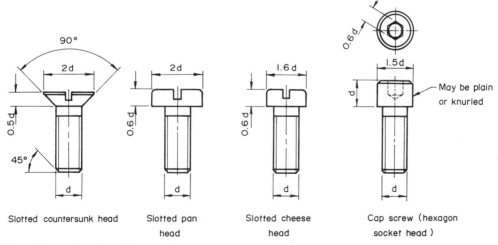

Figure 4.7 *Alternative head forms for screws*

Locking devices for screw fasteners

There is always a tendency for nuts fitted to vibrating machinery to slacken off gradually. This tendency must be resisted by a locking device. These are of two main types, frictional and positive, and some of the commoner examples of each are illustrated in Figure 4.8.

Lock nut

This frictional locking device is illustrated in Figure 4.8(a). A thin nut is first tightened down and a normal nut is screwed down on top of it. This nut is firmly held with a spanner and the thin nut slackened back slightly. The small amount of slackening wedges the nut threads against the bolt threads and jams the nuts in place. The top nut carries all the tensile load in the bolt so this should not be the thin nut. Often the thin nut is replaced by a normal nut.

Note that to jam the threads, some people prefer

to hold the thin nut and tighten the normal nut slightly.

Locking washers

Two forms of this very common frictional locking device are shown in Figure 4.8(b). Spring washers are wound from square or rectangular section wire. The single-coil type is illustrated but a double-coil type is also made.

When a nut is tightened down on a spring washer the washer thrusts the nut threads against the bolt threads, thus increasing the friction between them. The type shown has deflected ends at the split, producing sharp corners which dig into the nut and component and increase the locking effect. To prevent damage to the face of the component, a thin washer is generally used under the spring washer. Suitable drawing proportions for spring washers are 2d for the outside diameter and thickness 0.2d, where d is the major diameter of the fastener thread.

A variation of the spring washer is the star washer,

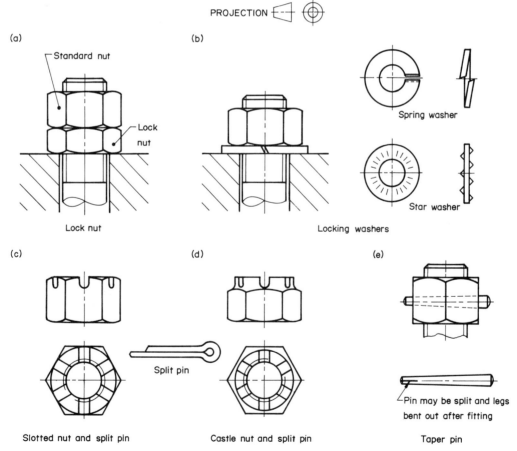

Figure 4.8 *Locking devices for screw fasteners*

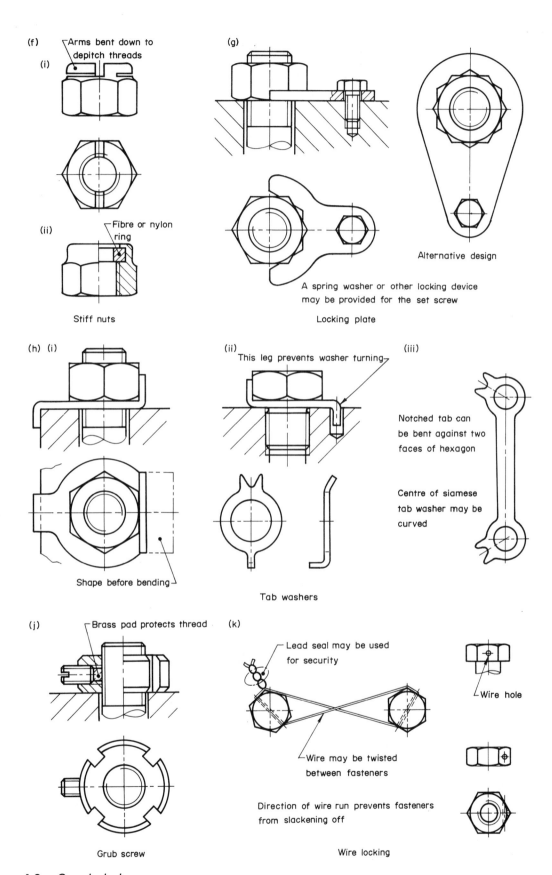

(f)
(i) Arms bent down to depitch threads

(ii) Fibre or nylon ring

Stiff nuts

(g)

Alternative design

A spring washer or other locking device may be provided for the set screw

Locking plate

(h) (i)

Shape before bending

(ii) This leg prevents washer turning

(iii) Notched tab can be bent against two faces of hexagon

Centre of siamese tab washer may be curved

Tab washers

(j) Brass pad protects thread

Grub screw

(k) Lead seal may be used for security

Wire may be twisted between fasteners

Direction of wire run prevents fasteners from slackening off

Wire hole

Wire locking

Figure 4.8 *Concluded*

of which there are several types. Internal or external projections on the faces of the washer form many sharp corners which dig into the component and the face of the nut.

Slotted nut

This is a hexagon nut with six slots cut across the top face, as shown in Figure 4.8(c). A split pin passes through one slot and a hole in the bolt, the legs of the pin being opened out and bent round the faces of the nut. This gives a positive lock.

It is sometimes difficult to line up a slot and the hole in the bolt, and the hole weakens the bolt. However, the height of the nut is increased compared with that of a normal nut, so that the slots do not weaken it.

Castle nut

Figure 4.8(d) shows this development of the slotted nut. The top of the nut carries a cylindrical rim into which six or eight slots are cut, depending on the size of the nut. A split pin passes through one slot and the hole in the bolt, and the legs of the pin are bent round the rim. With this nut the bent legs of the split pin are safer than with a slotted nut as they do not stand out on the nut faces.

Taper pin

Figure 4.8(e) shows a nut locked on to a bolt by a taper pin which passes through them. This is a very safe positive locking device, particularly if the small end of the taper pin is split and the legs are opened out.

A disadvantage of this locking device is that if the nut is removed it is very difficult to line up the holes in the nut and bolt again. The hole also weakens both parts.

Stiff nuts

Two examples of this frictional locking device are given in Figure 4.8(f). At (i) the cylindrical rim is slotted to form two arms which are deflected down slightly to de-pitch the threads in them. When the nut is fitted to a bolt the arms are forced up, producing friction between the nut and bolt threads.

At (ii) a collar with an internal diameter smaller than the thread minor diameter is fitted inside the cylindrical rim. When the nut is screwed on to the bolt, the bolt threads force their way through the collar which grips them tightly producing a frictional lock. The collar may be of fibre or nylon, the nylon being suitable for use at high temperatures.

A variation of the normal stiff nut is shown in Exercise 6.6. The nut is sawn halfway through and after assembly is locked by a hammer blow. This type of nut should only be used if it is not to be removed during the life of the assembly. Removal will damage the thread on the stud or bolt, which will have to be replaced.

Locking plates

Figure 4.8(g) shows two forms of this positive locking device, which is made from sheet metal. They have bi-hexagonal holes which engage with the hexagon on the nut. The plate is secured to the main part by a small set screw, which often has its own locking device. Locking plates may have a half hole, or they may enclose the nut completely.

Tab washers

Three common forms of this positive locking device are shown in Figure 4.8(h). At (i) is shown a type suitable for fitting over the edge of a flange. The washer is fitted under the nut, one leg is bent over the flange and the other up against a flat on the nut. If no flange is available, the form at (ii) can be used. Here a plug is shown locked by a tab washer having a small leg fitting in a hole drilled in the component, the other leg being bent up against the plug as before. This second leg has a vee cut in it, so that it can be bent against two faces of the hexagon if the final position of the plug requires it.

At (iii) is illustrated a siamese tab washer which is two tab washers made in one piece. Legs to prevent rotation of the washer are unnecessary. Siamese tab washers can be used only on faces with no projections between the bolt holes.

Grub screw

The circular nut in Figure 4.8(j) is tightened with a cee or claw spanner and locked by a grub screw. Tightening the grub screw presses a brass pad against the bolt threads giving a frictional lock. The pad prevents the end of the grub screw damaging the bolt threads.

Wire locking

Figure 4.8(k) shows two set bolts locked by this positive method. It is often used to lock plugs, pipe unions and nuts and similar parts on aero-engines.

A length of soft wire is passed through holes drilled in the bolt heads, pulled taut and the ends twisted together with a pair of pliers. Sometimes the wire is twisted together between the parts being locked. On nuts and pipe unions the holes are drilled through the corners of the hexagon.

Covers over components whose setting or position is fixed during manufacture often have wire locking through the securing screws with a lead seal fixed to the end. This prevents unauthorized alteration of the setting. The covers of domestic electricity meters are sealed in this way.

Self-assessment questions

4.11 Give a reason for using a plain washer under a nut.

4.12 How are screws with hexagon socket heads tightened?

4.13 Give the names of two frictional locking devices and two positive locking devices.

4.14 Is a star washer a frictional or positive locking device?

4.15 What is the difference between a castle nut and a slotted nut?

4.16 What advantage does a castle nut have over a slotted nut?

4.17 Some tab washers have a vee in the locking tab. What is the purpose of this vee?

4.18 Some tab washers have a small leg as well as a locking tab. What is the purpose of this leg?

4.19 Sketch a siamese tab washer.

4.20 When wire-locking a series of bolt heads on a pitch circle, in which direction should the wire run?

Exercise 4.8

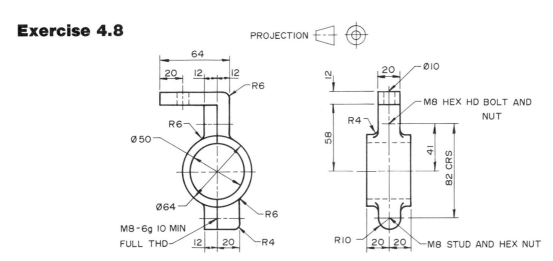

Draw the right-hand view of the pipe support with the screw fasteners and replace the other view with a section taken through the centre line.

Exercise 4.9

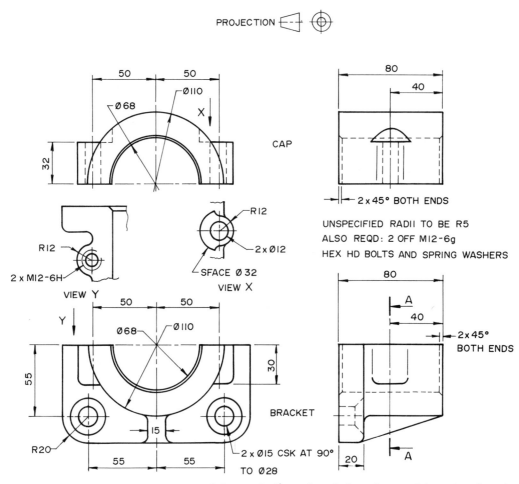

PROJECTION

CAP

Ø68
Ø110
X
50 50
32

80
40
2 x 45° BOTH ENDS

UNSPECIFIED RADII TO BE R5
ALSO REQD: 2 OFF M12-6g
HEX HD BOLTS AND SPRING WASHERS

R12
2 x M12-6H
VIEW Y

R12
2 x Ø12
SFACE Ø 32
VIEW X

Y
50 50
Ø68 Ø110
55
30
15
R20
55 55
2 x Ø15 CSK AT 90°
TO Ø28
BRACKET

80
A
40
2 x 45°
BOTH ENDS
20
A

Assemble the parts of the clamp and draw a half sectional elevation on AA and a plan view.

Exercise 4.10

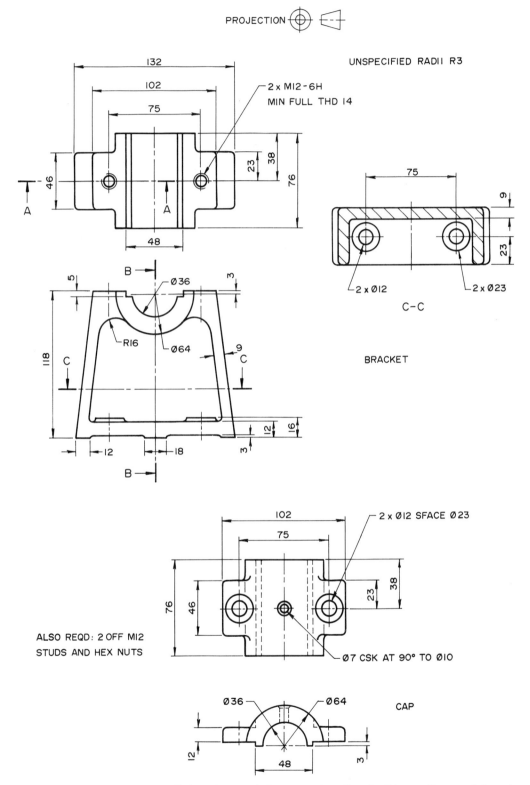

PROJECTION

UNSPECIFIED RADII R3

132
102
75

2 x M12-6H
MIN FULL THD 14

38
23
76
46

A
A

48

B
Ø36
5
3
R16
Ø64
118
C
C
9
12
16
12
18
3
B

BRACKET

75
9
23
2 x Ø12
2 x Ø23

C-C

102
75
2 x Ø12 SFACE Ø23
76
46
23
38

ALSO REQD: 2 OFF M12
STUDS AND HEX NUTS

Ø7 CSK AT 90° TO Ø10

Ø36
Ø64
CAP
12
48
3

Assemble the parts of the split bearing and draw an elevation half in section on AA and an end view in section on BB.

Exercise 4.11

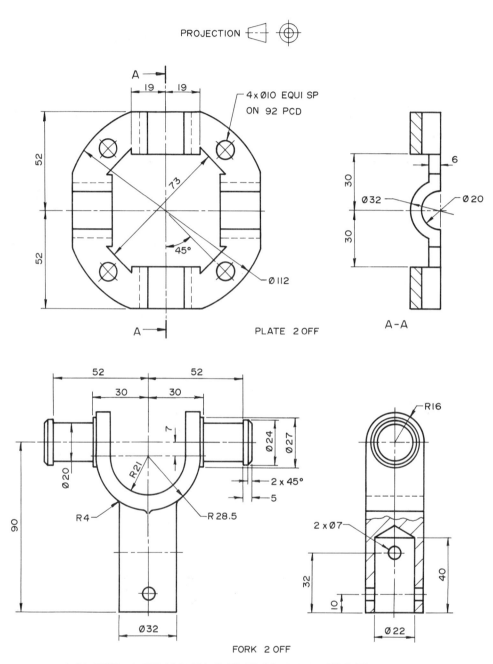

PROJECTION

4 x Ø10 EQUI SP ON 92 PCD

19 | 19

52

73

45°

52

Ø112

A — A

PLATE 2 OFF

A-A

30

30

Ø32

Ø20

6

52 | 52

30 | 30

7

Ø24

Ø27

Ø20

R21

R4

R28.5

90

Ø32

2 x 45°

5

R16

2 x Ø7

32

10

40

Ø22

FORK 2 OFF

ALSO REQD: 4 OFF MIO HEX HD BOLTS 24 LG AND HEX NUTS

Assemble the parts for the Hooke's joint or universal coupling and draw an elevation looking on the joint line between the plates and an end view.

Post test 4

1 For the bolt shown write down the dimensions which represent:
   (a) the major diameter of the thread;
   (b) the minor diameter of the thread;
   (c) the bolt length;
   (d) the thread length.

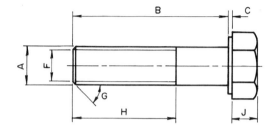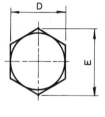

2 A thread is designated M24 × 2 – 6H. State whether the thread is internal or external and whether it is coarse or fine pitch.

3 Complete the given views with an M20 blind tapping having a full form thread depth of 40.

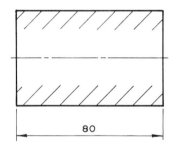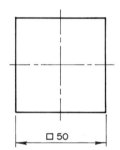

4 Make sketches of and name two positive and two frictional locking devices for screwed fasteners. State which is a positive and which a frictional device.

5

Fastenings

Principal learning objective

The student uses the conventions in British Standards for representing fastenings.

Riveted joints

A riveted joint is a permanent fastening, that is, it cannot be dismantled without destroying the rivets.

Riveted joints may be divided into two types: those for heavy-gauge work, such as in boiler manufacture; and those for light-gauge work, such as in aircraft manufacture. Heavy-gauge work also includes shipbuilding and structural steel work, but welded joints are increasingly being used in these applications. Welded joints are dealt with in Chapter 6.

Rivet heads for heavy-gauge work

Figure 5.1 shows some types of rivet head used in heavy-gauge work. The rivets are made with a single head. When the plates are joined the rivets are made red hot, inserted in the holes in the plates and a second head is formed from the tail of the rivet by pneumatic or hydraulic pressure. The rivet contracts as it cools and this draws the plates closely together. Rivet holes are a little bigger than the rivet diameter, but when the rivets are closed they fill the holes.

Types of riveted joints

Figure 5.2 illustrates the two types of joint. The **lap joint** connects plates which lie in parallel planes and overlap each other. The **butt joint** connects plates lying in the same plane, the ends of the main plates being covered by one or two **straps** or **cover plates**. Figure 5.2(b) and (c) illustrates single and double cover plate joints.

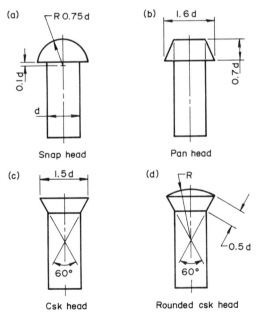

Figure 5.1 *Types of rivet heads for heavy-gauge work*

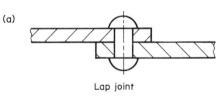
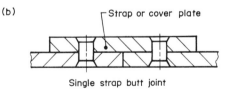
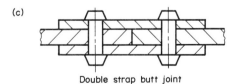

Figure 5.2 *Types of riveted joints*

Rivet arrangements

Figure 5.3 shows how rivets may be arranged in the joint. In **chain riveting** the rivets are placed directly opposite each other. In **diagonal** or **zig-zag** riveting the rivets are staggered from row to row.

Riveted joints are said to be **single**, **double** or **triple riveted**, etc. This refers to the number of rows of rivets counted from the joint end of one plate. So a single riveted lap joint has a one row of rivets, as shown in Figure 5.2(a). A single riveted butt joint has two, one in each main plate, as shown in Figure 5.2(b) and (c).

(a)

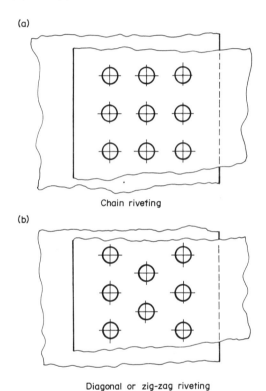

Chain riveting

(b)

Diagonal or zig-zag riveting

Figure 5.3 *Rivet arrangements*

Riveted joints for light-gauge work

Three commonly used rivet heads for this class of work are illustrated in Figure 5.4. The rivets fall into three classes: **solid**, **semi-tubular** and **tubular**, as shown in Figure 5.5. The solid rivet is the most commonly used and is made in a wide range of materials, diameters and head shapes. Solid rivets are frequently closed with a **bumped-up head**, illustrated in Figure 5.4(b), the head being formed with hammer blows. Semi-tubular rivets have the advantage that the force needed to close or 'set' them is less than that for a solid rivet. Tubular or hollow rivets can be used in blind positions, that is,

in positions where only one side of the work is accessible.

Examples of tubular rivets

Two types are shown in Figure 5.6, both of which are set by drawing a mandrel through the rivet bore. With the Tucker 'Pop' rivet, illustrated in Figure 5.6(a), the mandrel breaks at a previously weakened point when the rivet has been set. The mandrel in the Avdel 'Briv' rivet shown in Figure 5.6(b), does not break and is pulled through the rivet bore intact and then discarded.

(a)

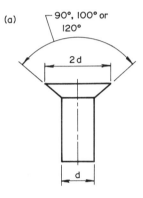

Countersunk head

(b)

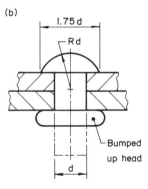

Snap head

(c)

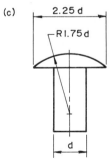

Mushroom head

Figure 5.4 *Types of rivet heads for light-gauge work*

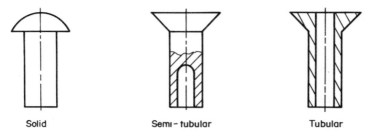

Solid   Semi-tubular   Tubular

Figure 5.5   *Types of rivets for light-gauge work*

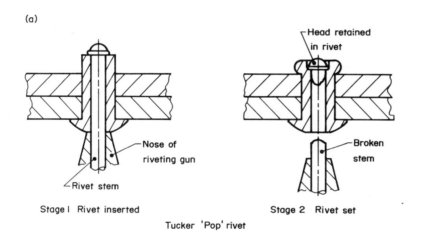

(a)

Head retained
in rivet

Nose of
riveting gun

Broken
stem

Rivet stem

Stage I   Rivet inserted   Stage 2   Rivet set

Tucker 'Pop' rivet

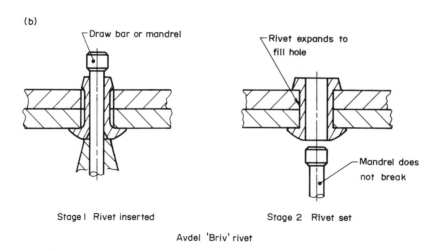

(b)

Draw bar or mandrel

Rivet expands to
fill hole

Mandrel does
not break

Stage I   Rivet inserted   Stage 2   Rivet set

Avdel 'Briv' rivet

Figure 5.6   *Examples of tubular rivets*

Self-assessment questions

5.1 Sketch three types of rivet head used in heavy-gauge work.

5.2 Sketch sectional views through a double riveted lap joint and a double riveted, single cover plate, butt joint with countersunk head rivets. Both joints are diagonally riveted. Add a second view to show this.

5.3 Sketch a sectional view through a light-gauge joint with a snap head rivet. Show the rivet closed with a bumped-up head.

5.4 Which types of rivet for light-gauge work can be used where only one side of the joint is accessible?

5.5 Sketch a section through a joint which is secured by a Tucker 'Pop' rivet.

Keys

A key is inserted axially between a shaft and a hub to prevent relative rotation of the parts. Examples of various types of key are given in Figure 5.7.

Parallel keys

These are shown in Figure 5.7(a). The cross section of the key is square or rectangular and the top and bottom faces are parallel. The key is sunk half in the shaft and half in the hub, measured at the edges of the keyways. It fits closely in the sides of the keyway and has clearance at the top.

Taper keys

Square and rectangular keys are also made with a taper of 1 : 100 on the top face. This makes the key more secure in the keyways. A **gib head** may be added, as shown in Figure 5.7(b). This allows the key to be driven in by tapping the sloping face of the head and withdrawn by tapping a tapered drift between the hub and the back of the gib head.

Feather keys

A parallel key may be used as a **feather**, that is, the hub may slide along the shaft while rotating with it, as with the gears in a gearbox. The feather is often fixed to the shaft with countersunk screws. A feather key is illustrated in Figure 5.7(c).

Woodruff keys

A Woodruff key, shown in Figure 5.7(d), is in the form of a segment of a circle. It fits in a corresponding recess in the shaft. It adjusts itself to any taper in the hub keyway and is widely used on machine tools. It cannot be used as a feather because it may jam, and it has the disadvantage that the deep keyway weakens the shaft.

Round keys

Two types of round key are shown in Figure 5.7(e). The plain type is circular and is a driving fit in a hole drilled half in the shaft and half in the hub. Round keys may also be threaded and screwed in position, in which case they usually have a square head for fitting which is machined off when the key is in place.

Saddle keys

These are of two types, **flat** and **hollow**, and are illustrated in Figure 5.7(f). They are suitable for use only when the torque to be transmitted is low, and the hollow type is used solely for temporary fastenings.

Dimensioning of keyways

The recommended ways of dimensioning the different types of keyways are shown in Figure 5.8. These dimensions can be checked by practical inspection procedures.

(a)

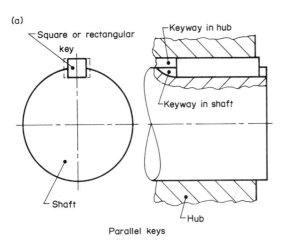

Square or rectangular key

Keyway in hub

Keyway in shaft

Shaft

Hub

Parallel keys

(b)

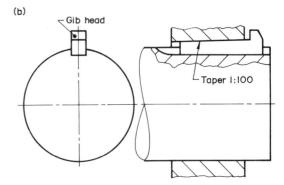

Gib head

Taper 1:100

Gib head allows key to be withdrawn using a tapered drift
Taper keys also made without gib heads in square and
rectangular forms

Taper keys

(c)

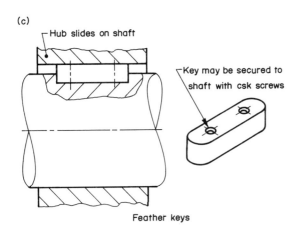

Hub slides on shaft

Key may be secured to
shaft with csk screws

Feather keys

(d)

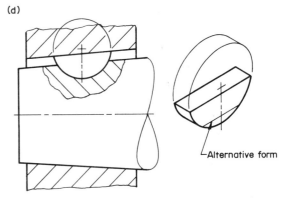

Alternative form

Woodruff keys may jam if used with parallel shafts

Woodruff keys

(e)

Key half in shaft, half in hub

Plain

Square end removed after
key is fitted

Threaded

Round keys

(f)

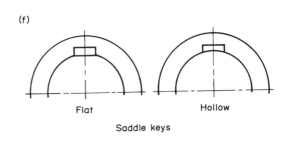

Flat

Hollow

Saddle keys

Figure 5.7 *Types of keys*

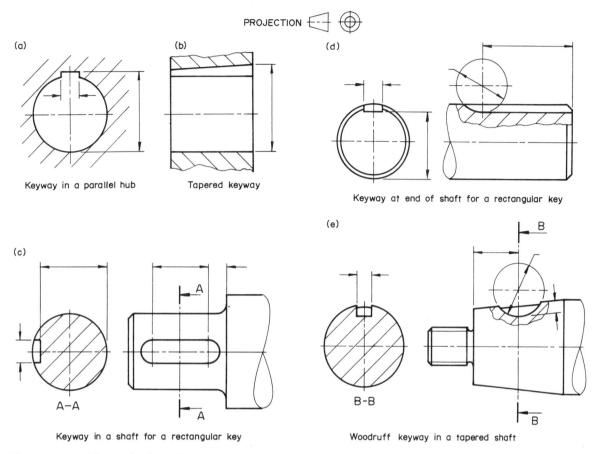

PROJECTION

(a)
Keyway in a parallel hub

(b)
Tapered keyway

(c)
A–A
Keyway in a shaft for a rectangular key

(d)
Keyway at end of shaft for a rectangular key

(e)
B–B
Woodruff keyway in a tapered shaft

Figure 5.8   *Dimensioning of keyways*

Dowel pins

Screw fasteners hold parts together and generally are designed to take tensile loads only. They are usually unsuitable for locating parts transversely, thus resisting shear loads, because the clearance between the fastener and its mating hole is often too great. To locate the parts transversely **dowel pins**, as shown in Figure 5.9, are often used.

To ensure that the parts are located relative to each other as accurately as possible, the dowel holes are drilled together. The circular dowel pins are an interference fit in one part and a close clearance fit in the other, so the holes need to be reamed separately. Figure 5.9 also shows the convention of filling in opposite quadrants of the circular view of the dowel pin. This conventional representation is widely used in drawing offices.

Circlips or spring retaining rings

Where parts such as bearings or collars need to be

positioned along the length of a shaft, **circlips** or **spring retaining rings**, as illustrated in Figure 5.10, are often used. However, the circlips are a relatively loose fit in their grooves, so the axial location is not very accurate. Note that the correct pin-ended pliers must be used to fit and remove the circlips.

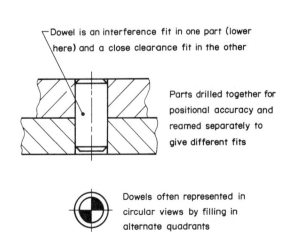

Dowel is an interference fit in one part (lower here) and a close clearance fit in the other

Parts drilled together for positional accuracy and reamed separately to give different fits

Dowels often represented in circular views by filling in alternate quadrants

Figure 5.9   *Location of parts using dowels*

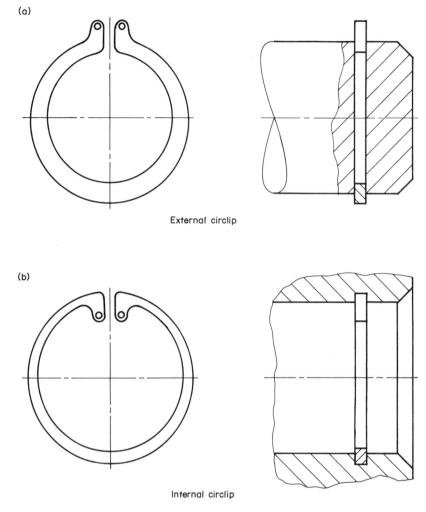

(a)

External circlip

(b)

Internal circlip

Holes in lugs enable pin-ended pliers to spring circlips for fitting and removal

Figure 5.10 *Circlips (spring retaining rings)*

Self-assessment questions

5.6   What is the purpose of fitting a key between a hub and a shaft?

5.7   Sketch a rectangular taper key with a gib head. Which face of the key is tapered and by how much?

5.8   State one advantage and one disadvantage of the Woodruff key.

5.9   State the types of load which screw fasteners and dowels are designed to take.

5.10   When fitting dowels, why are the parts drilled together?

Post test 5

1 Identify the types of rivet head shown.

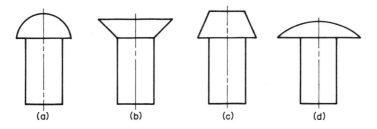

(a)   (b)   (c)   (d)

2 For the riveted joint shown:
   (a) zigzag riveting has been used.   True or false?
   (b) the joint is a lap joint.   True or false?
   (c) the joint is double riveted.   True or false?

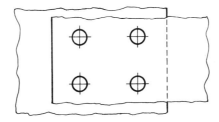

3 Sketch a view of a shaft which is secured to a hub by a taper key with a gib head. State:
   (a) the purpose of the gib head;
   (b) whether a taper key can be used as a feather. Give reasons for your answer.

4 Copy the two views of the shaft end. Omit the given dimensions and replace them with the correct ones.

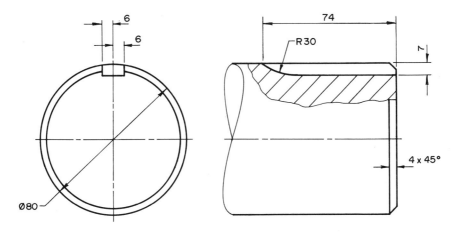

Conventional practices and representations

Principal learning objective

The student uses the conventions in British Standards for representing knurling, squares on shafts, related holes, bearings, springs, gears, machining symbols and welding symbols.

Introduction

The cost of a drawing, represented by the time spent to make it, is part of the total cost of the component, and any means of keeping it down should be used. One way of doing this is by using conventional practices and conventional representations of common features.

The primary object of using conventional practices and representations is to save time in the production of drawings, but some also save space on the drawing. This may mean that a view can be drawn to a larger scale, thus improving the readability of the drawing. On the other hand, space saved may mean that a smaller size of drawing sheet can be used. The smaller the drawing sheet, the cheaper is its storage, transmission and reproduction. Various ways of achieving both these objects are recommended in BS 308: Part 1, and those which occur most frequently are discussed below.

Conventional practices

Symmetry

Many components are symmetrical about a centre line or axis and can often be represented satisfactorily by a half view. This is illustrated in Figure 6.1(a). For parts which are symmetrical about two centre lines at right angles a quarter view may be used. An example is shown in Figure 6.1(b).

To show that a half or quarter view has been drawn, two short, thin, parallel lines are drawn across the symmetry demarcation line at each end. These symmetry symbols are at right angles to the symmetry demarcation line. To emphasise further that a half or quarter view is shown, the outlines of the part extend slightly beyond the line of symmetry.

Sometimes parts which are basically symmetrical have asymmetrical features. The symmetry convention may still be used in such cases, provided that the half view which shows the asymmetrical features is drawn and that the asymmetrical features are identified by a note. Figure 6.1(c) is an example.

It may be sufficient to show an adequate segment of a circular part, as shown in Figure 6.1(d). However, all uses of the symmetry convention should be considered carefully to avoid any ambiguities or misunderstanding of the drawing. A drawing which is not clear will waste time and may result in the production of scrap.

Enlarged part views

On occasion, components have features which, in the general scale of the drawing, are too small to be dimensioned clearly. Such features may be enclosed in a thin circle, to which an identifying letter is added. An enlarged part view is drawn, carrying the same identification letter and its scale, and this part view is used for the dimensions. Figure 6.2 illustrates this convention.

Repetitive information

On some drawings identical parts, such as nuts and bolts, and identical features, such as holes, appear

(a)

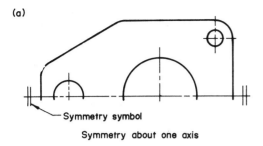

Symmetry symbol

Symmetry about one axis

(b)

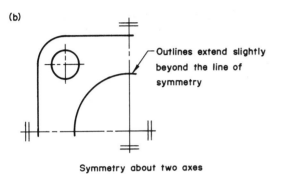

Outlines extend slightly beyond the line of symmetry

Symmetry about two axes

(c)

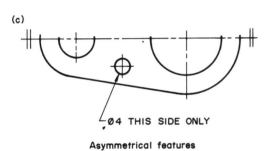

Ø4 THIS SIDE ONLY

Asymmetrical features

(d)

PROJECTION

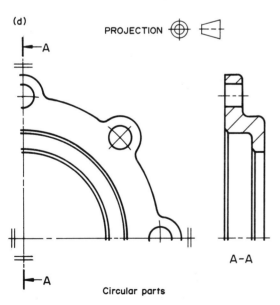

A–A

Circular parts

Do not use this convention if the drawing could be misunderstood

Figure 6.1 *Symmetrical objects*

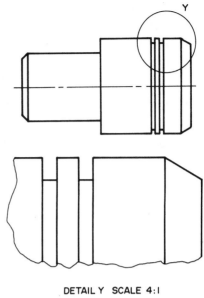

DETAIL Y SCALE 4:1

Note that the irregular boundary is a *thin* line

Figure 6.2 *Enlarged part view for clear dimensioning*

(a)

Repeated identical parts or features

(b)

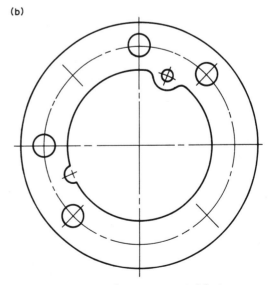

Single features adjacent to repeated features

Figure 6.3 *Repeated information*

many times. Repeated illustrations of them may be avoided by drawing one, and indicating the positions of the others by their centre lines, as shown in Figure 6.3(a).

Figure 6.3(b) shows a part which has eight identical holes. Adjacent to two pairs of these holes are single features. The holes on each side of the single features should be shown in full.

Self-assessment questions

6.1 What two objectives may be achieved by the use of conventional practices and representations on drawings?

6.2 For what objects could the use of quarter views be considered?

6.3 The symmetry convention can be used on views of objects which are asymmetrical.   True or false?

Conventional representations of common features

Interrupted views

These may be used to show long parts of constant cross section. A piece from the centre of the part is imagined to be removed and the ends of the view are drawn close together. Figure 6.4(a), (b) and (c) shows break lines used for different cross sections. Figure 6.4(d) shows a break line which can easily be drawn by machine. Break lines are thin. Sometimes a revolved section is drawn in the gap in the interrupted view. See Figure 2.9.

Knurling

The circumferential surface of the heads of adjusting screws is often roughened by knurling to provide a better finger grip. Two types, **straight** and **diamond knurling**, are in common use and the conventional representations for both are illustrated in Figure 6.5. These use thin lines. The spacing of the lines should be quite wide to avoid them closing up on reduced-size prints. Straight knurling is often used on the heads of hexagon socket cap screws.

Rolling bearings

If ball and roller bearings are shown fully in section, the races, cages and balls and rollers would all have to be drawn. Since rolling bearings are not dimensioned on an assembly drawing, but are identified by a note, such detail is unnecessary. Instead, the conventional representation in Figure 6.6 is used.

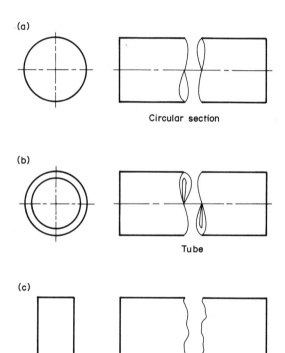

(a)

Circular section

(b)

Tube

(c)

Other sections

(d)

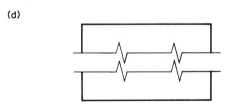

This break line may replace any of those above.

It is particularly useful for drawings made by machine

Figure 6.4 *Interrupted views*

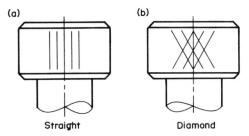

Straight   Diamond

The thin lines representing the knurling are widely spaced and shown on part of the treated surface only

Figure 6.5 *Types of knurling*

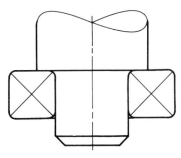

Any type of ball and roller bearing in section may be represented using this convention

Figure 6.6 *Rolling bearings*

Flats and squares on shafts

These features are represented conventionally by thin lines drawn diagonally on the flat surfaces, as shown in Figure 6.7.

Gears

The conventional representation for single gears and racks is shown in Figure 6.8. Teeth on a circular view may be omitted and represented by a thick circle for the tips of the teeth and a thin chain circle for the pitch circle. No circle is shown through the roots of the teeth.

On a sectional view, hatching is omitted from the teeth, even if the cutting plane passes through them. The pitch line is shown as a thin chain line and the root surface by a thick line.

For a rack, shown in Figure 6.8(d), one or two teeth are drawn at each end and the remainder shown by a thick line representing the tips and a thin line representing the roots. The pitch line is shown by a thin chain line. The dimensions p and q indicate how the teeth are located on the blank.

Figure 6.9 shows the conventional representation of spur and bevel gears, and worm and wormwheels in mesh. Note that for bevel gears the pitch cones are shown by thin chain lines.

Cylindrical helical springs

The helices on these springs should be shown conventionally as straight lines. The pitch of the coils need not be exact and only two or three coils should be shown at each end, as illustrated in Figure 6.10. The remaining coils are indicated by a thin chain line passing through the centre of the cross section of the wire of the spring.

For diagrams and schematic drawings the representation may be further simplified to a thick single line as shown. In this case the cross section of the spring material may be indicated in words or by a symbol as in Figure 6.10(a). Here the diameter

PROJECTION

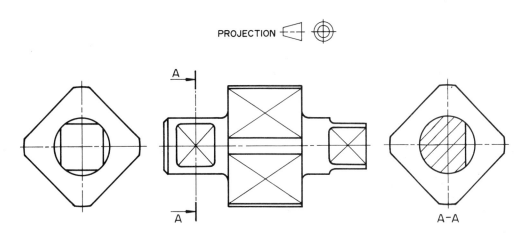

Figure 6.7 *Flats and squares on shafts*

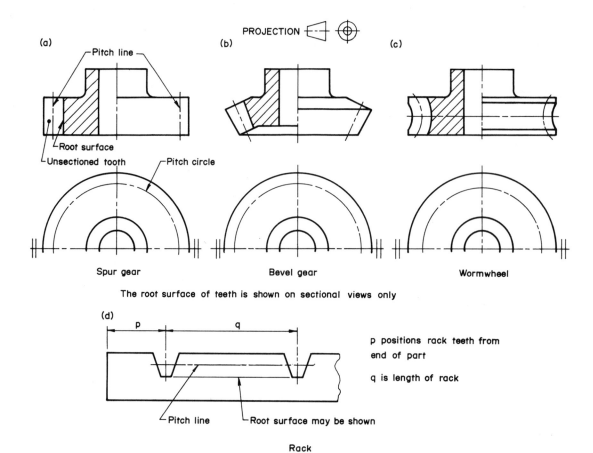

PROJECTION ⊏ ⊕

(a)

Pitch line

Root surface

Unsectioned tooth

Pitch circle

Spur gear

(b)

Bevel gear

(c)

Wormwheel

The root surface of teeth is shown on sectional views only

(d)

p   q

Pitch line   Root surface may be shown

Rack

p positions rack teeth from end of part

q is length of rack

Figure 6.8   *Conventional representation for gears*

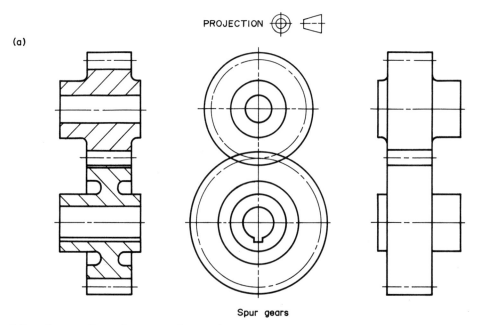

PROJECTION ⊕ ⊏

(a)

Spur gears

Figure 6.9   *Conventions for gears in mesh*

(b)

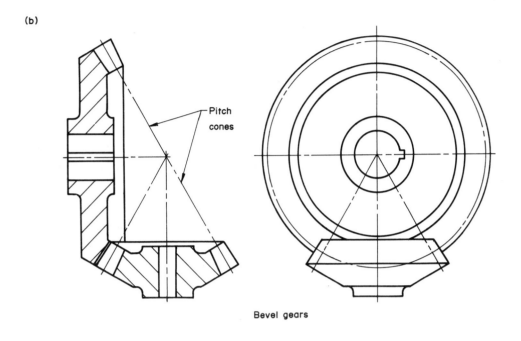

Pitch
cones

Bevel gears

(c)

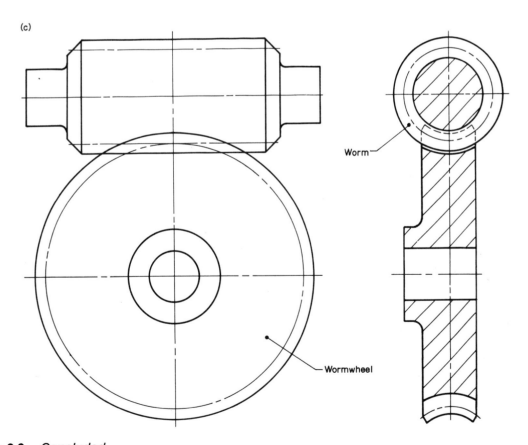

Worm

Wormwheel

Figure 6.9 *Concluded*

symbol is used because the cross section of the wire is circular. If the cross section were square, the square symbol would be used.

The slope of the coils depends on the hand of the helix and is opposite on outside views and sections. The springs in Figure 6.10 are all right-handed. Where necessary the drawing should indicate 'wound left- (or right-) hand'.

(a)

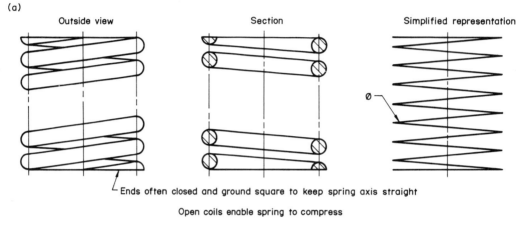

Outside view   Section   Simplified representation

Ø

Ends often closed and ground square to keep spring axis straight

Open coils enable spring to compress

Cylindrical helical compression spring. Wire of circular cross section

(b)

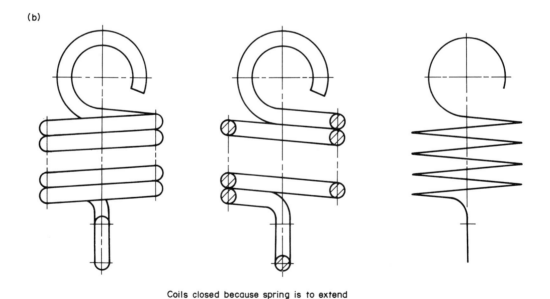

Coils closed because spring is to extend

Cylindrical helical tension spring. Wire of circular cross section

Figure 6.10 *Conventional representations for cylindrical helical springs*

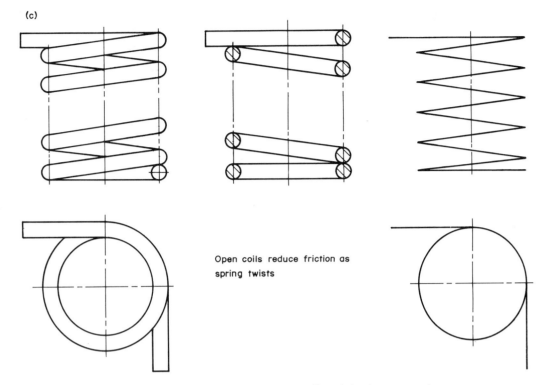

(c)

Open coils reduce friction as
spring twists

Cylindrical helical torsion spring. Wire of circular cross section

Figure 6.10   *Concluded*

Self-assessment questions

6.4   Sketch an undimensioned interrupted view of a solid shaft 150 long and 25 diameter.

6.5   A spur gear with 16 teeth is 30 wide with a diameter over the tips of the teeth of 90 and a pitch circle diameter of 78. The depth of the teeth is 13. The gear has a bore of 25. Using all appropriate conventional practices and representations, draw the following undimensioned views of the gear:

(a)   a half sectional view,

(b)   a circular view.

6.6   In what way are the conventional representations similar for a rolling bearing in section and a squared end to a shaft?

6.7   Draw the outside view and sectional view of a cylindrical helical compression spring with an outside diameter of 70 and a wire diameter of 8. The spring is 100 long. The ends of the spring are closed and ground square.

Machining and surface texture symbols

The basic symbol is a tick as shown in Figure 6.11(a). If machining of the surface is required, a bar is added to the basic symbol, as in Figure 6.11(b), but if removal of material from the surface is not permitted, a circle is added to the basic symbol as shown in Figure 6.11(c).

Surface texture values in micrometres (μm) are

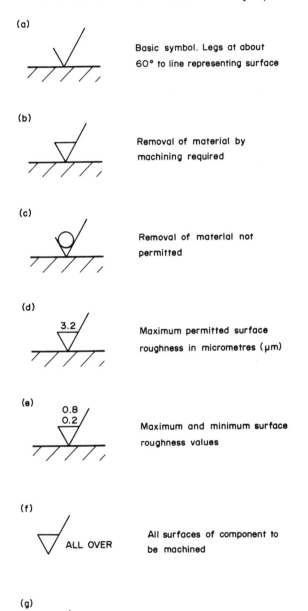

<table>
<tr><td>(a)</td><td>Basic symbol. Legs at about 60° to line representing surface</td></tr>
<tr><td>(b)</td><td>Removal of material by machining required</td></tr>
<tr><td>(c)</td><td>Removal of material not permitted</td></tr>
<tr><td>(d) 3.2</td><td>Maximum permitted surface roughness in micrometres (μm)</td></tr>
<tr><td>(e) 0.8 0.2</td><td>Maximum and minimum surface roughness values</td></tr>
<tr><td>(f) ALL OVER</td><td>All surfaces of component to be machined</td></tr>
<tr><td>(g) 3.2 EXCEPT AS STATED</td><td>General value of surface roughness for most surfaces of component</td></tr>
</table>

Figure 6.11 *Indications of machining and surface texture*

added to the machining symbol as required. In Figure 6.11(d) the maximum permitted surface roughness is specified; when both the maximum and minimum values of surface roughness need to be specified, the maximum value is placed above the minimum, as in Figure 6.11(e).

When all the surfaces of a component are to be machined, the indication shown in Figure 6.11(f) is added to the drawing, usually near the title block. A surface texture value may be added if necessary.

If the same surface texture value applies to a number of surfaces of the component, the symbol and value may be stated in a note, as shown in Figure 6.11(g). The note usually appears near the title block.

Symbols should, in general, be shown once on each surface, preferably on the same view as the size and location dimensions of the surface. Where space is restricted the symbols may be shown adjacent to the dimension on an extension of the dimension line; or on a leader line or extension line, see Figure 6.12. In each case the symbol should be normal to the line on which it is placed.

Values added to the symbols should be positioned so that they can be read from the bottom or from the right-hand side of the drawing, as in Figure 6.12. The symbol or arrow may point from outside the material of the part, either to a line representing the surface or to an extension of such a line.

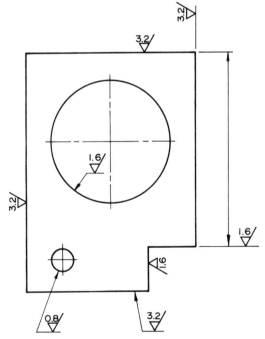

Figure 6.12 *Application of machining symbols and surface texture values*

Welding symbols

Welding is used extensively for fastening parts together permanently, instead of using bolts, screws, rivets or other fasteners. It is also used to fabricate machine parts and other structures which otherwise would have been cast or forged, and in structural engineering and ship building.

There are numerous different welded joints, a selection of which is shown in Figure 6.13. To avoid having to draw the actual shape of welds, each type is represented by a symbol, illustrated in Figure 6.13.

The symbol for the type of weld is applied to the single-line representation of the joint by a leader line which has a horizontal extension. The extension is called the **reference line**, see Figure 6.14.

The weld symbol is placed below the reference line if the external surface of the weld is on the arrow side of the joint, see Figure 6.14(a), but if the weld is to be on the other side of the joint, as in Figure 6.14(b), the weld symbol is placed above the reference line.

If the weld is to be on both the arrow side and the other side of the joint, weld symbols are placed above and below the reference line.

For welds which are made within the plane of the joint, as in Figure 6.14(c), the weld symbol is placed across the reference line.

When a joint is shown by a single line on a drawing, the arrow side of the joint is regarded as the side nearest to the reader of the drawing, following the conventions of orthographic projection.

When a weld is made all round a joint (a **peripheral weld**) a circle is placed where the leader line joins the reference line, as shown in Figure 6.15.

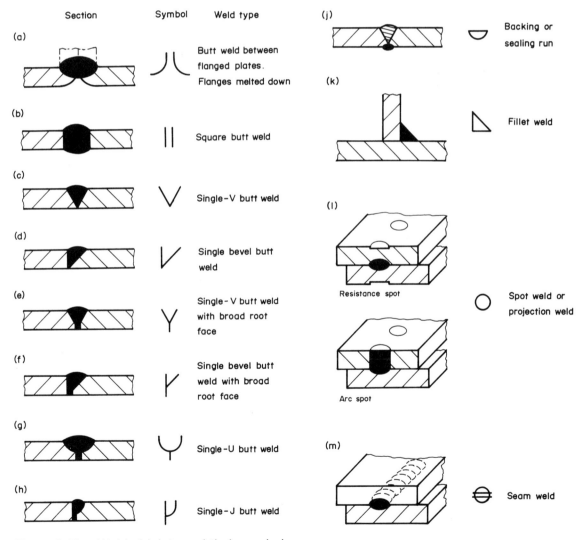

Figure 6.13 *Welded joints and their symbols*

101

PROJECTION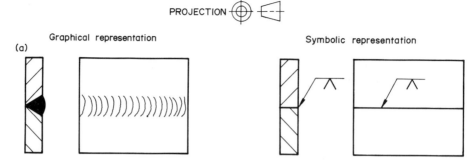

(a)

Graphical representation   Symbolic representation

Symbol is below the reference line if the external surface of the
weld (weld face) is on the arrow side of the joint

(b)

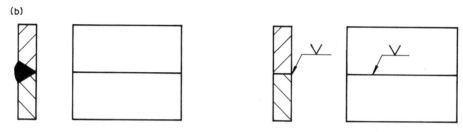

Symbol is above the reference line if the external surface of
the weld (weld face) is on the other side of the joint

(c)

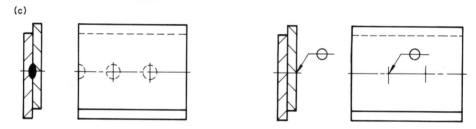

Symbol is across the reference line when welds are made within
the plane of the joint

Figure 6.14   *Position of the welding symbol*

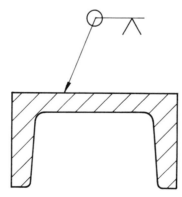

Figure 6.15   *Symbol for a peripheral weld*

Self-assessment questions

6.8 Sketch the machining symbol which is used when machining of a surface is not permitted.

6.9 After matching, the surface texture value of a surface is to be between $0.8\,\mu m$ and $1.6\,\mu m$. Sketch the machining symbol.

6.10 All the surfaces of a part are to be machined with a maximum surface texture value of $3.2\,\mu m$. Sketch the symbol.

6.11 Sketch the symbols for the following welds:
- (a) square butt weld,
- (b) backing run,
- (c) spot weld,
- (d) single bevel butt weld with broad root face.

6.12 Sketch the leader line, reference line and weld symbol for:
- (a) fillet weld on the arrow side of a joint,
- (b) fillet weld on the other side of a joint,
- (c) fillet weld on both sides of a joint,
- (d) seam weld.

Exercise 6.1

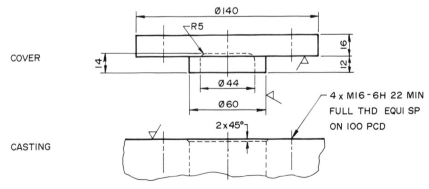

ALSO REQD: 4 OFF MI6-6g SET BOLTS AND SPRING WASHERS

Assemble the parts and draw a half sectional elevation and a half plan view.

Exercise 6.2

PROJECTION

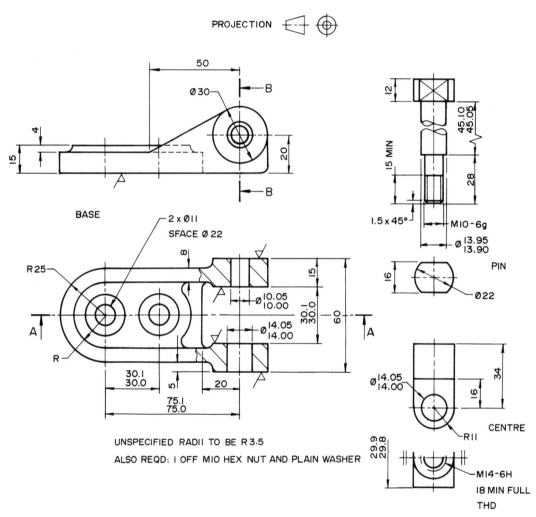

50

Ø 30

B

4

15

20

B

BASE

2 x Ø11
SFACE Ø 22

R 25

8

15

R

Ø 10.05 / 10.00

30.1 / 30.0

60

Ø 14.05 / 14.00

A

A

30.1 / 30.0

5

20

75.1 / 75.0

UNSPECIFIED RADII TO BE R 3.5

ALSO REQD: I OFF MIO HEX NUT AND PLAIN WASHER

12

45.10 / 45.05

15 MIN

28

1.5 x 45°

MIO - 6g

Ø 13.95 / 13.90

PIN

16

Ø 22

Ø 14.05 / 14.00

34

16

CENTRE

R11

29.9 / 29.8

M14 - 6H

18 MIN FULL
THD

Assemble the parts and draw a sectional elevation on AA and a sectional end view on BB.

Exercise 6.3

PROJECTION

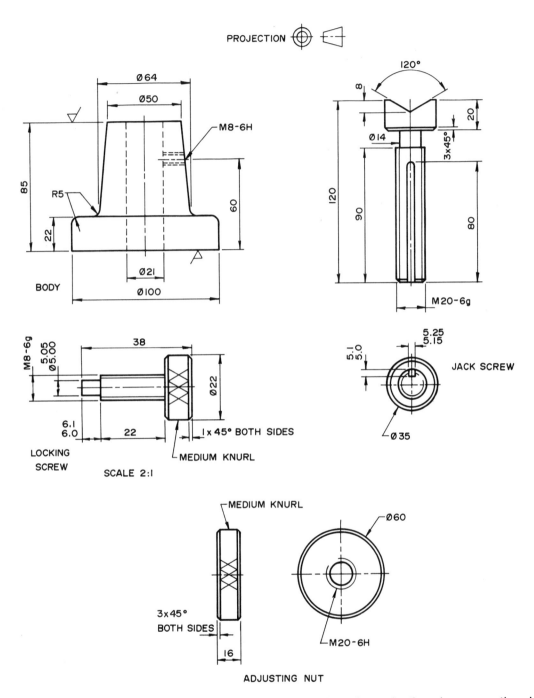

BODY

JACK SCREW

LOCKING SCREW

SCALE 2:1

MEDIUM KNURL

M20-6H

ADJUSTING NUT

Assemble the parts of the screw jack and using third angle projection draw a sectional elevation corresponding to the given view of the body and a plan. Break out a section around the slot in the jack screw.

Exercise 6.4

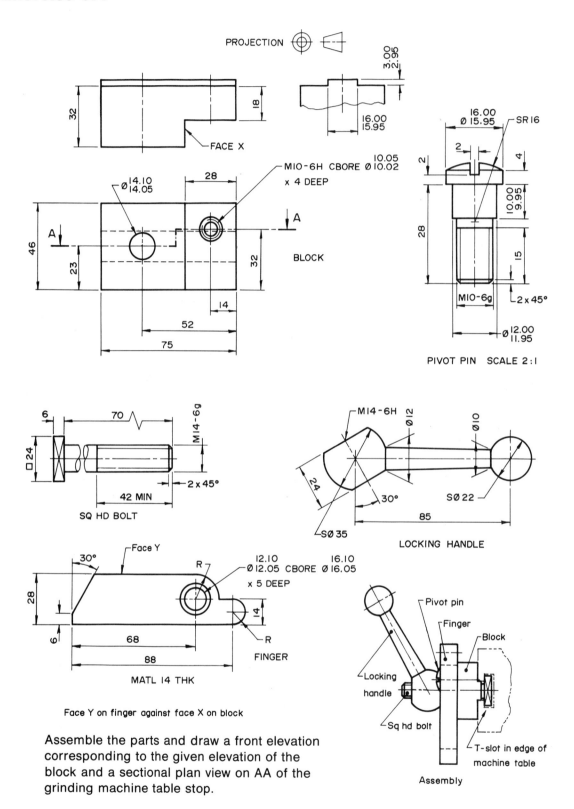

PROJECTION

FACE X

Ø 14.10 / 14.05

M10-6H CBORE Ø 10.05 / 10.02 x 4 DEEP

BLOCK

PIVOT PIN   SCALE 2:1

SR 16

16.00 / Ø 15.95

M10-6g

2 x 45°

Ø 12.00 / 11.95

SQ HD BOLT

M14-6g

2 x 45°

42 MIN

LOCKING HANDLE

M14-6H

Ø 12

Ø 10

SØ 22

SØ 35

FINGER

Face Y

12.10 / Ø 12.05 CBORE Ø 16.10 / 16.05 x 5 DEEP

R

R

MATL 14 THK

Face Y on finger against face X on block

Assembly

Pivot pin

Finger

Block

Locking handle

Sq hd bolt

T-slot in edge of machine table

Assemble the parts and draw a front elevation corresponding to the given elevation of the block and a sectional plan view on AA of the grinding machine table stop.

Exercise 6.5

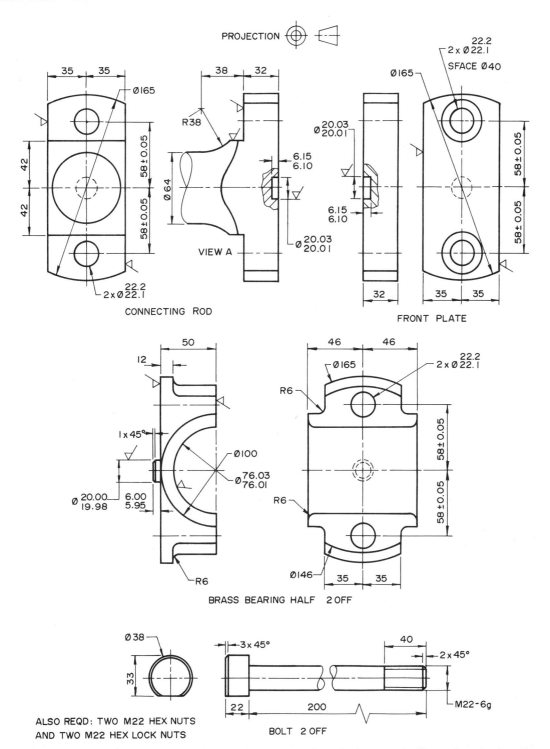

PROJECTION

CONNECTING ROD

VIEW A

FRONT PLATE

BRASS BEARING HALF 2 OFF

ALSO REQD: TWO M22 HEX NUTS
AND TWO M22 HEX LOCK NUTS

BOLT 2 OFF

Assemble the parts of the connecting rod end and draw a half sectional elevation corresponding to view A of the rod and an end view looking on the hexagon nuts.

Exercise 6.6

PROJECTION

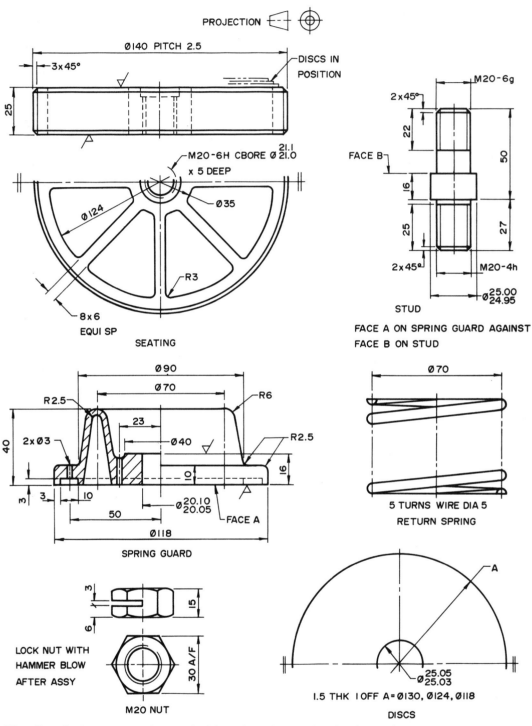

Ø140 PITCH 2.5

DISCS IN POSITION

3x45°

25

M20-6H CBORE Ø 21.0 21.1 x 5 DEEP

Ø35

Ø124

R3

8x6 EQUI SP

SEATING

M20-6g

2x45°

22

FACE B

16

25

50

27

2x45°

M20-4h

Ø 25.00 24.95

STUD

FACE A ON SPRING GUARD AGAINST FACE B ON STUD

Ø90

Ø70

R6

R2.5

23

Ø40

R2.5

40

2x Ø3

10

16

3

3

10

Ø 20.10 20.05

FACE A

50

Ø118

SPRING GUARD

Ø70

5 TURNS WIRE DIA 5 RETURN SPRING

3

15

6

LOCK NUT WITH HAMMER BLOW AFTER ASSY

30 A/F

M20 NUT

A

Ø 25.05 25.03

1.5 THK 1OFF A=Ø130, Ø124, Ø118

DISCS

The discs in the pump valve are held against the seating by the spring in the spring guard. Pressure in the pump moves the discs upwards and opens the valve. Assemble the parts and draw a sectional elevation corresponding to the given elevation of the seating and a half plan.

Exercise 6.7

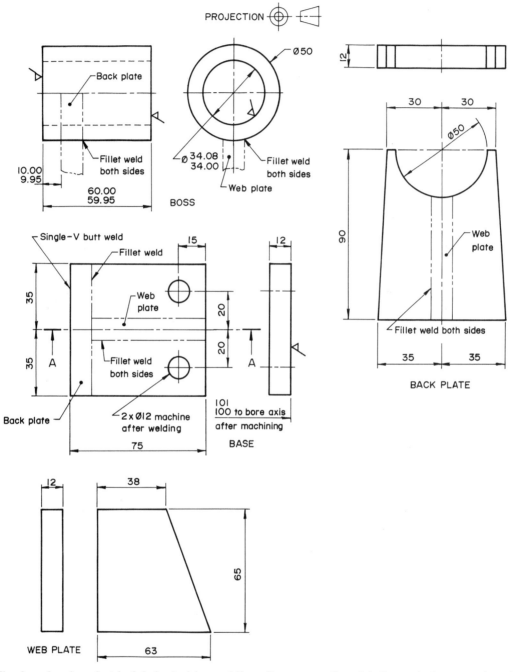

PROJECTION

Back plate

Fillet weld both sides

10.00
9.95

60.00
59.95

BOSS

Ø50

Ø 34.08
34.00

Fillet weld both sides

Web plate

12

30   30

Ø50

90

Web plate

Fillet weld both sides

35   35

BACK PLATE

Single-V butt weld

Fillet weld

Web plate

15

12

35

20

Web plate

20

35   A

Fillet weld both sides

A

Back plate

2 x Ø12 machine after welding

101
100 to bore axis
after machining

BASE

75

12

38

65

WEB PLATE   63

The bearing bracket is fabricated by welding. Draw a section AA through the completed bracket and add an end view looking on the edge of the web plate. Identify the welds by using the correct symbols and show the dimensions needed for the machining after welding.

Exercise 6.8

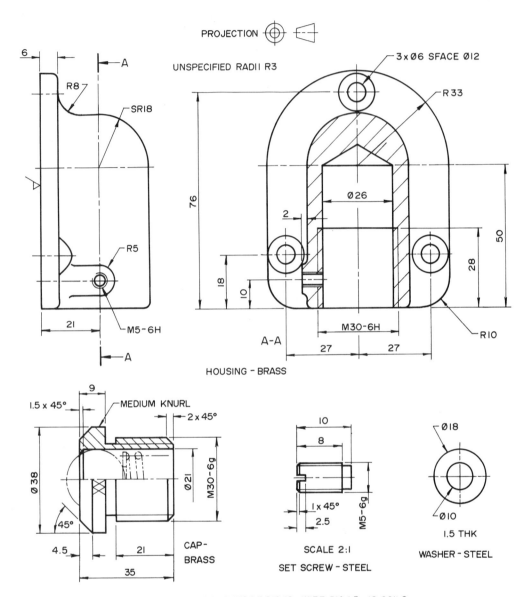

PROJECTION

UNSPECIFIED RADII R3

6

R8

SR18

76

R5

21

M5-6H

A-A

A

3 x Ø6 SFACE Ø12

R33

Ø26

2

18

10

50

28

M30-6H

27   27

R10

HOUSING – BRASS

9

1.5 x 45°

MEDIUM KNURL

2 x 45°

Ø38

Ø21

M30-6g

45°

4.5

21

35

CAP-
BRASS

10

8

1 x 45°

2.5

M5-6g

SCALE 2:1

SET SCREW – STEEL

Ø18

Ø10

1.5 THK

WASHER – STEEL

SPRING PARTICULARS: OUTSIDE DIA 18, WIRE DIA 1.5, 12 COILS,
FREE LENGTH 75, ENDS CLOSED AND GROUND SQUARE

ALSO REQD: 21 DIA STEEL BALL HARDENED AND GROUND

Assemble the parts of the door catch and draw twice full size a sectional view replacing the
given left-hand view of the housing and an outside view replacing section AA.

Post test 6

1 (a) Objects which are symmetrical about one axis can be represented by half views.   True or false?

   (b) Identical parts which appear many times on a view must all be shown for completeness. True or false?

   (c) On a sectioned view of a gear hatching is shown on the gear teeth.   True or false?

   (d) A flat on a shaft is represented by thin diagonal lines drawn on a true shape view of the flat. True or false?

   (e) On an assembly drawing a spring can be represented by a single thick line.   True or false?

2 State the meaning of the given machining symbols.

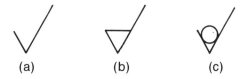

        (a)   (b)   (c)

3 All the surfaces of a part are to be machined and have a surface roughness of 0.8 μm. Show how this could be indicated on the drawing.

4 State the meaning of the given welding symbols.

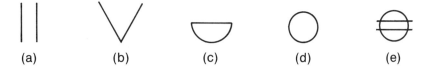

   (a)   (b)   (c)   (d)   (e)

5 Draw the symbolic representation for a single bevel butt joint where the weld face is on the arrow side of the joint.

6 Make drawings in section of the welded joints represented by the given symbols. All plates are 15 thick.

    (a)   (b)   (c)   (d)

7

Pictorial projections

Principal learning objectives

The student produces neat and clear drawings using isometric and oblique projection. The student draws from given orthographic drawings isometric views (without using the isometric scale) and oblique views which include rectilinear and curved objects.

Introduction

Two or more views of an object drawn in orthographic projection describe its size and form accurately. Each view, however, shows only two of the three dimensions, length, breadth and height, of the object, and the views have to be arranged on the paper in a particular way, as explained in Chapter 1. So, to produce correct drawings and to visualise objects correctly from them, a thorough understanding of the projection system is essential.

For some purposes, drawings are needed that are proportioned accurately and can be understood easily without a knowledge of orthographic projection. Such drawings can be produced using several methods of projecting a single view of an object on a plane. The single view shows three faces of the object, more or less as an observer would see them, and is called a **pictorial drawing**.

Although pictorial drawings have the advantage that they do not need an understanding of orthographic projection to be made or understood, they do have disadvantages which limit their use. These include:

(a) Views produced by some methods are so distorted as to be of no use.
(b) Pictorial drawings generally take longer to make than drawings in orthographic projection. Those which most closely represent what an observer sees take the longest time and are the most complicated to draw.
(c) Pictorial drawings are difficult to dimension.
(d) In some methods circles and curves take a long time to draw.
(e) Some lines in pictorial drawings cannot be measured.

In spite of these disadvantages pictorial drawings are used extensively in instruction and repair manuals, parts catalogues and sales literature. They are also useful for supplementing and clarifying orthographic drawings of complex parts, and to sketch ideas during the preliminary stages of a new design.

Of the methods of producing a pictorial drawing which are available, only two are covered here: **isometric** and **oblique**.

Self-assessment question

7.1 List the advantages and disadvantages of pictorial drawings compared with drawings made in orthographic projection.

Isometric drawing

The basic principles which underlie the making of an isometric drawing are illustrated in Figure 7.1.

Figure 7.1(a) shows a cube positioned so that a face is parallel to a vertical plane. This is the projection plane for the isometric view. Orthogonal projectors from the cube produce a square view on the projection plane, which is a true view of one face

of the cube. In Figure 7.1(b) the plan view of the cube has been turned through 45°. Two faces of the cube are now visible when it is projected orthogonally onto the projection plane.

In Figure 7.1(c) the **body diagonal** of the cube is perpendicular to the projection plane. The body diagonal is the line which passes through the cube from one corner to the opposite corner. With the

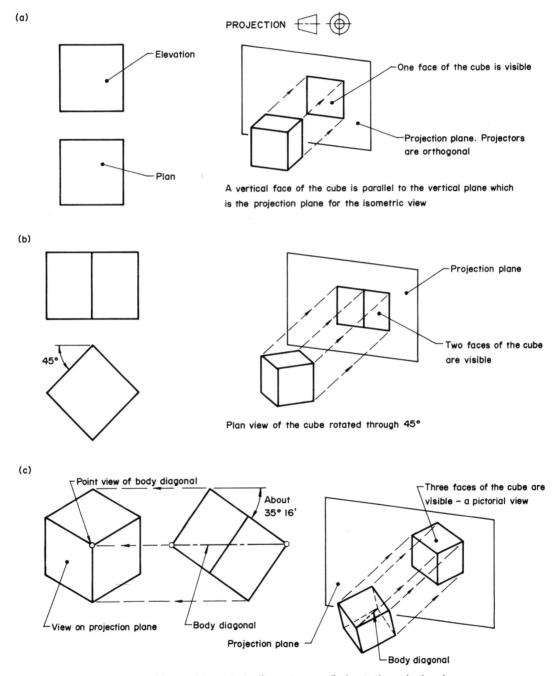

Figure 7.1 *The isometric cube*

cube in this position its orthogonal projection on the plane shows three of its faces. Therefore this projection of the cube is a pictorial view.

When the cube is in the position shown in Figure 7.1(c), all of its edges are at the same angle to the projection plane. Therefore its orthogonal projection on the vertical plane will show all the edges as the same length, that is, they will be equally foreshortened. From this fact comes the name of the system, **iso** – equal, **metric** – measure.

The projected view in Figure 7.1(c) has been drawn again in Figure 7.2(a). It is found that all the edges of the cube appear at 30° to the horizontal, except AD, OC and BE which appear vertical. Thus, edges which are parallel on the cube or other object being drawn, remain parallel in the isometric view. This makes the view easy to draw, but since no account is taken of perspective it is slightly distorted. However, the distortion becomes objectionable only when very long objects are drawn.

The lines OA, OB and OC, redrawn in Figure

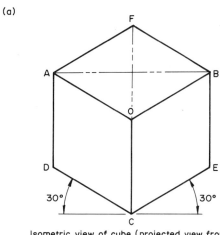

(a)

Isometric view of cube (projected view from figure 7.1)

OA, OB and OC are the isometric axes. OADC, OBEC and OAFB are the isometric planes

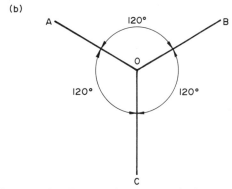

(b)

Figure 7.2   *Isometric axes and planes*

7.2(b), are called the **isometric axes**, and on them the drawing is built up. Alternative positions of the axes may be used, as will be seen later, these positions resulting from different orientations of the object to the projection plane. The 120° angles between the axes remain the same, though, whichever position is used. Lines parallel to the axes are called **isometric lines**; others are **non-isometric lines**. The planes OADC, OAFB and OBEC are the **isometric planes**.

When preparing an isometric drawing, measurements may be scaled directly only along isometric lines or parallel to them. This may be verified by comparing the diagonals AB and FO in Figure 7.2(a). These lines, which are non-isometric lines, have the same true length, but their isometric lengths are obviously not equal. Note, however, that AB, which is parallel to the projection plane, is in fact the true length of both of them.

Objects composed entirely of isometric lines

These objects are drawn easily, as all the measurements in the orthographic views can be scaled directly on to the isometric view. It is unnecessary to draw the orthographic views. Figure 7.3 is an example of this type of object. It illustrates the convenient technique of first drawing a box which will just contain the object, and then building up the shape of the object inside the box. Most objects can be treated in this way, and fewer errors are made by beginners who use it.

Note that hidden detail is omitted from isometric and other pictorial drawings unless its use is essential to clarify the shape of the object. Non-essential hidden detail is often confusing and the view takes longer to draw.

Objects with non-isometric lines

Lines on an object which are located by angles are **non-isometric** lines. Angles cannot be laid off directly on an isometric drawing as they do not appear as their true sizes. That this is so can be seen from Figure 7.2(a). All the angles on the cube are right angles but they are represented by angles of 60° or 120°, never by a right angle.

Lines positioned by angles are drawn by fixing their ends with ordinates that are isometric lines. The part of the orthographic view which shows the line located by the angle is drawn, and the ordinates transferred to the isometric view. The method is shown in Figure 7.4.

114

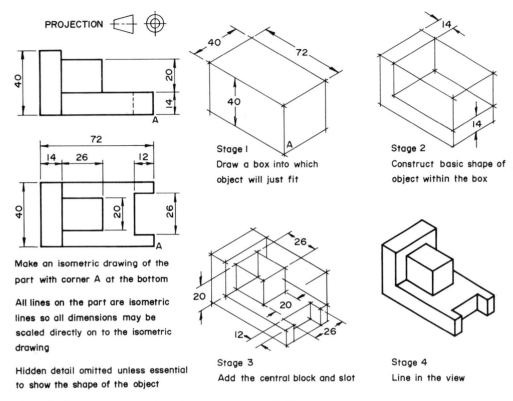

PROJECTION

Stage I
Draw a box into which
object will just fit

Stage 2
Construct basic shape of
object within the box

Make an isometric drawing of the
part with corner A at the bottom

All lines on the part are isometric
lines so all dimensions may be
scaled directly on to the isometric
drawing

Hidden detail omitted unless essential
to show the shape of the object

Stage 3
Add the central block and slot

Stage 4
Line in the view

Figure 7.3 *Objects composed entirely of isometric lines*

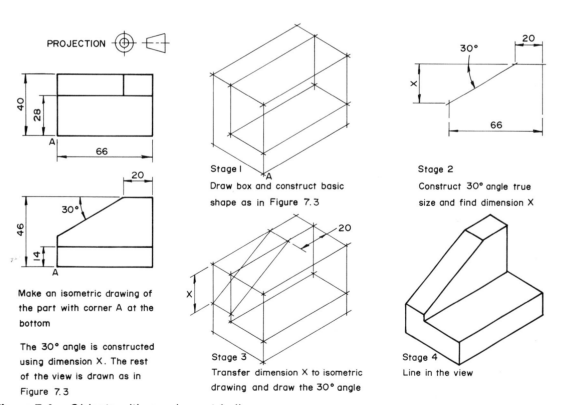

PROJECTION

Stage I
Draw box and construct basic
shape as in Figure 7.3

Stage 2
Construct 30° angle true
size and find dimension X

Make an isometric drawing of
the part with corner A at the
bottom

The 30° angle is constructed
using dimension X. The rest
of the view is drawn as in
Figure 7.3

Stage 3
Transfer dimension X to isometric
drawing and draw the 30° angle

Stage 4
Line in the view

Figure 7.4 *Objects with non-isometric lines*

Make isometric drawings of the given objects, positioned so that the corner marked A is the lowest point on the drawing. Do not draw the given orthographic views.

Exercise 7.1

Exercise 7.2

PROJECTION

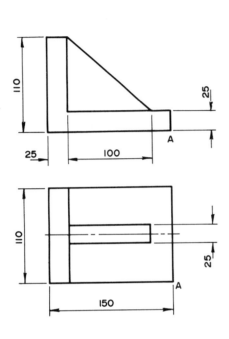

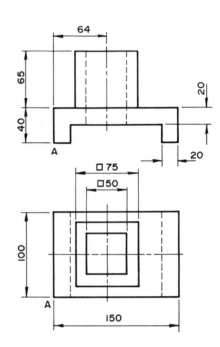

Exercise 7.3

Exercise 7.4

PROJECTION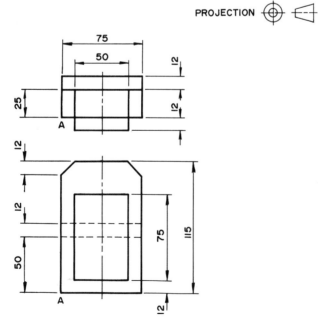

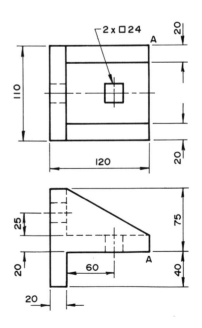

Make isometric drawings of the given objects, positioned so that the corner marked A is the lowest point on the drawing. Do not draw complete orthographic views.

Exercise 7.5

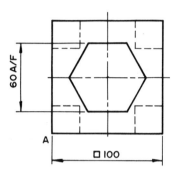

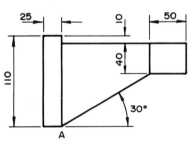

□100

60 A/F

60

25

12

50 | 2 SLOTS

A

Exercise 7.6

PROJECTION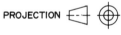

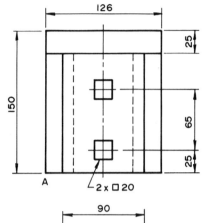

126

25

150

65

25

A

2 x □ 20

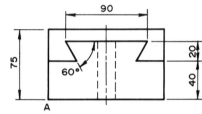

90

75

60°

20

40

A

Exercise 7.7

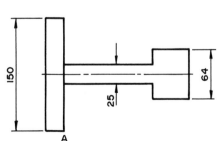

25

10

50

40

110

30°

A

150

64

25

A

Exercise 7.8

PROJECTION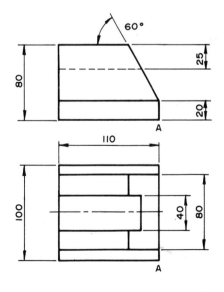

60°

80

25

20

A

110

100

40

80

A

117

Objects with a curved profile

These are drawn by selecting a number of points on the curved profile in the orthographic view, and transferring ordinates of each point to the isometric view, as shown in Figure 7.5. Usually a parallel curve is required. This can be obtained by drawing parallel lines through the points on the first curve in the appropriate direction, and marking off on them the width of the detail. Too many points should not be taken or the work becomes tedious, but too few points will give an inaccurate profile.

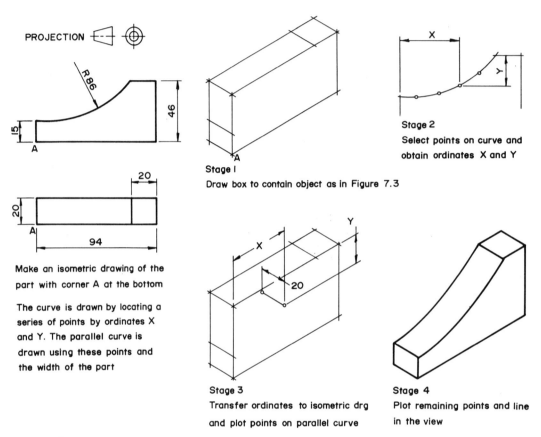

PROJECTION

R 86

46

15

A

20

20

A

94

Make an isometric drawing of the part with corner A at the bottom

The curve is drawn by locating a series of points by ordinates X and Y. The parallel curve is drawn using these points and the width of the part

Stage 1
Draw box to contain object as in Figure 7.3

X

Y

Stage 2
Select points on curve and obtain ordinates X and Y

X

Y

20

Stage 3
Transfer ordinates to isometric drg and plot points on parallel curve

Stage 4
Plot remaining points and line in the view

Figure 7.5   *Objects with a curved profile*

Self-assessment questions

7.2   What is the angle between adjacent isometric axes?

7.3   What technique can be used to help the drawing of an isometric view?

7.4   When drawing an isometric view of a curve, the ordinates to points on the curve are isometric lines.   True or false?

7.5   Why is hidden detail usually omitted from isometric drawings?

Circles in isometric drawing

Construction by ordinates. Any circle on an isometric plane will appear as an ellipse which may be drawn in several ways. Ordinates may be used as for drawing a curved profile, and this method is illustrated in Figure 7.6. From one quadrant of the full size circle, dimensions X and Y are transferred to the isometric view. The ellipse is completed with a french curve.

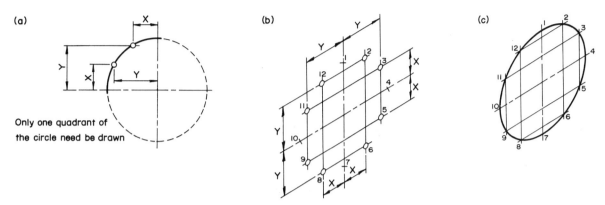

Figure 7.6 *Circle construction by ordinates*

Exercise 7.9

Exercise 7.10

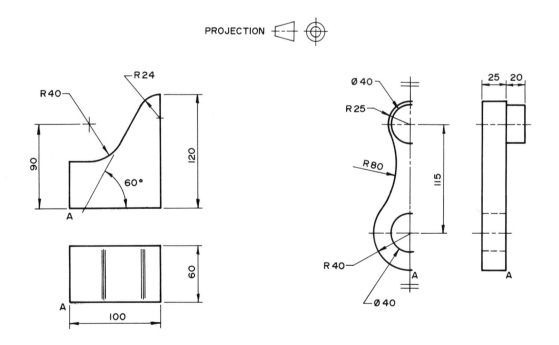

PROJECTION

Make isometric drawings of the given objects with corner A as the lowest point. Use the ordinate method to construct the isometric circles and curves.

Construction by the circumscribing square.
The circle in Figure 7.7(a) has been circumscribed by a square with a diagonal added. The square is redrawn in the isometric view and eight points on the ellipse are obtained from the intersection of the centre lines and square, and the diagonals and circle. See Figure 7.7(b). The ellipse is completed as before with a french curve.

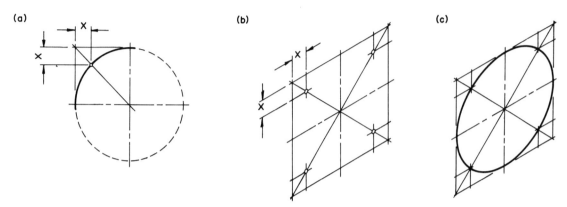

Figure 7.7 *Circle construction by circumscribing square method*

Exercise 7.11   Exercise 7.12

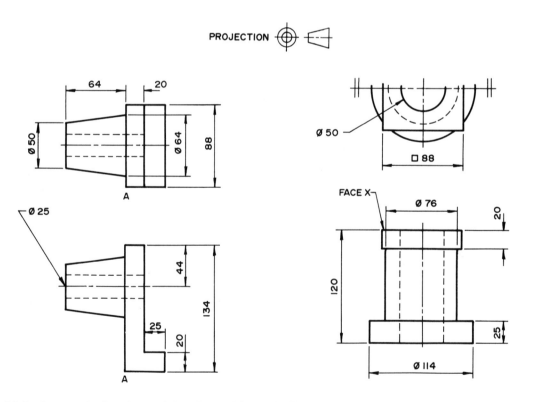

PROJECTION

Make isometric drawings of the given objects. In Exercise 7.11 corner A is to be the lowest point of the view. In Exercise 7.12 face X is to be at the top of the view. Draw the isometric circles using the circumscribing square method.

Approximate method. The ellipse occurs so frequently on isometric drawings that an approximate method using circular arcs is often used to draw it. One such method, called the **four-centre method**, is shown in Figure 7.8(a).

The centre lines EF and GH are first drawn and their intersection made the centre of an isometric square ABCD. The length of the sides of the square is equal to the diameter of the required circle. The long diagonal AC of the square is drawn, and either B or D is joined to the mid-points of the opposite sides. Where these lines cross the long diagonal are two arc centres. The other centres are the corners B and D of the isometric square.

The construction is completed more quickly if it is noted that all construction lines for the arc centres on the long diagonal are either horizontal or at 60° to the horizontal.

When drawing a cylinder as in Figure 7.8(b), or a hole the far end of which is visible, the construction in Figure 7.8(a) need be carried out for only one end. To obtain the centres for the other, lines parallel to the cylinder axis are drawn through centres A, B and C, and on them the length L of the cylinder is marked off. This fixes the centres X, Y and Z.

If a semicircle or quadrant is needed, only enough of the construction should be drawn to find the required centre or centres. For other fractions of a circle the ordinate method must be used, since the approximate method deviates considerably from a true ellipse at several points on the curve. This means that the approximate curve will not blend with other curves which are tangential to it at these points.

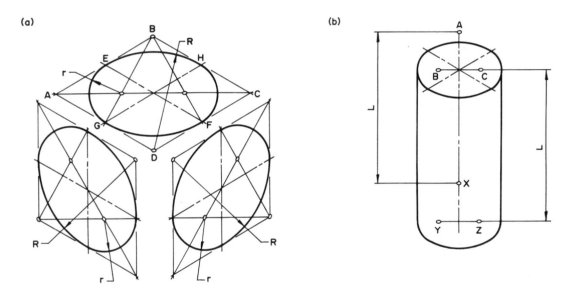

Figure 7.8 *Approximate (four-centre) circle construction*

Self-assessment questions

7.6 Any circle on an isometric plane will appear as an ellipse in an isometric view.   True or false?

7.7 A true ellipse can be drawn using the four-centre method.   True or false?

7.8 When using the circumscribing square method to draw an ellipse in an isometric view, the sides of the square must be isometric lines.   True or false?

Exercise 7.13   # Exercise 7.14

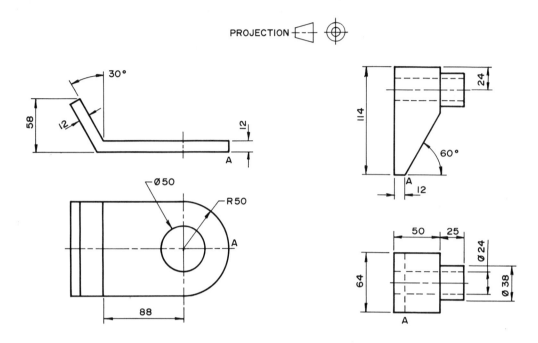

Make isometric drawings of the given objects with corner A as the lowest point. Use the approximate construction to draw the isometric circles.

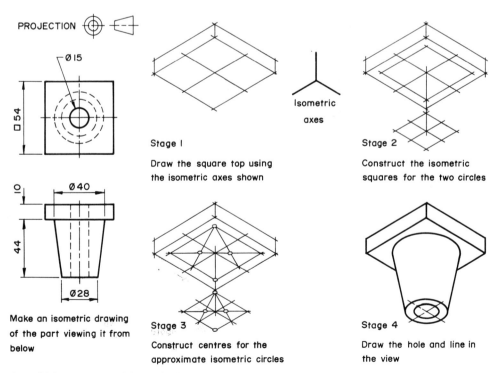

PROJECTION

Ø 15

□ 54

Make an isometric drawing of the part viewing it from below

Ø 40

10

44

Ø 28

Stage I

Draw the square top using the isometric axes shown

Isometric axes

Stage 2

Construct the isometric squares for the two circles

Stage 3

Construct centres for the approximate isometric circles

Stage 4

Draw the hole and line in the view

Figure 7.9 *Objects viewed from below*

122

Ellipse templates. There is no doubt that the time taken to draw isometric circles is a serious disadvantage of the projection system. When a large number of circles is required on the drawing the work becomes tedious in the extreme, whichever method described above is used. In these circumstances isometric ellipse templates are by far the simplest way of drawing the isometric circles, since the only construction needed is to lay out the centre lines of the circles. These are aligned with the corresponding centre lines on the template and the ellipse traced round.

It is obviously impossible to provide templates for every possible ellipse, but by choosing a template which is a little too large and leaning the pen or pencil outwards from the hole, a satisfactory curve can usually be drawn. Alternatively the template may be shifted slightly as each quarter of the ellipse is traced out.

Alternative positions of the isometric axes

Some parts are shown to their best advantage by varying the position of the isometric axes. In Figure 7.9 the object when in use is viewed from below, and the position of the axes is chosen to produce a view as seen from below. Occasionally long objects may be best drawn with their axes horizontal, as in Figure 7.10.

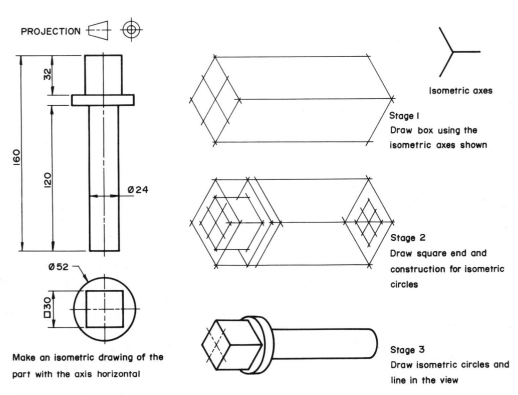

Figure 7.10 *Objects with axis shown horizontal*

Exercise 7.15

Exercise 7.16

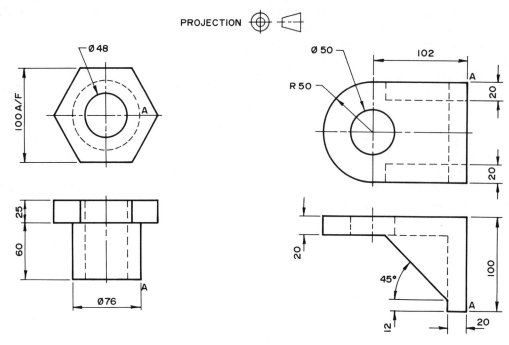

PROJECTION

Make isometric drawings of the given objects viewing them from below, with corner A as the lowest point. Draw the isometric circles using the approximate construction.

Oblique projection

It was stated above that the time taken to draw isometric circles was a serious disadvantage of the projection system. With oblique projection this disadvantage often disappears and views in this projection system can usually be drawn much quicker than in isometric projection.

The principles of the system are shown in Figure 7.11. In Figure 7.11(a) the elevation and plan of an object are shown in orthographic projection. In Figure 7.11(b) a new elevation is projected from the plan using projectors at 45° to the projection plane. This means that all edges of the object which are perpendicular to the projection plane project as true lengths, as do all heights on the object. It follows then that all faces of the object which are perpendicular or parallel to the projection plane project as true shapes.

In Figure 7.11(b) the new elevation shows two faces of the object, whereas for a pictorial view three faces must be visible. Figure 7.11(c) shows how the projection of the third face is obtained. The projectors at 45° to the projection plane are tipped down at some angle, which is often 30°. The top face of the object then becomes visible and the resulting pictorial view is an **oblique projection**.

In oblique projection the object is generally positioned with a face parallel to the projection plane. Such faces project as their true shapes, which means that all their edges are true lengths. Faces perpendicular to the projection plane do not project as true shapes, but again, all their edges are true lengths. These facts mean that, in general, an oblique projection of an object may be drawn directly, without first drawing the orthographic views. Three axes, corresponding to b^3g^3, b^3c^3 and b^3h^3 are drawn and on them the view is built up. The axis b^3h^3 is called the **cross** or **receding axis**.

Faces on the object which are parallel to the projection plane project as true shapes. Therefore if faces which include circles and circular arcs are positioned parallel to the projection plane, the curves may be drawn with compasses. This saves time in drawing the oblique projection.

(a)

Elevation

Orthographic views of object

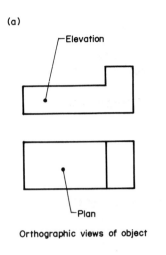

(b)

New elevation shows two faces of the object

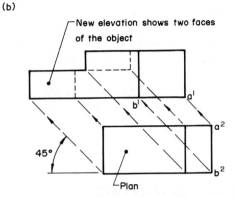

45°

When projectors from plan are at 45° to projection plane, edges perpendicular to projection plane, such as AB, project as TL. All heights remain TL

(c)

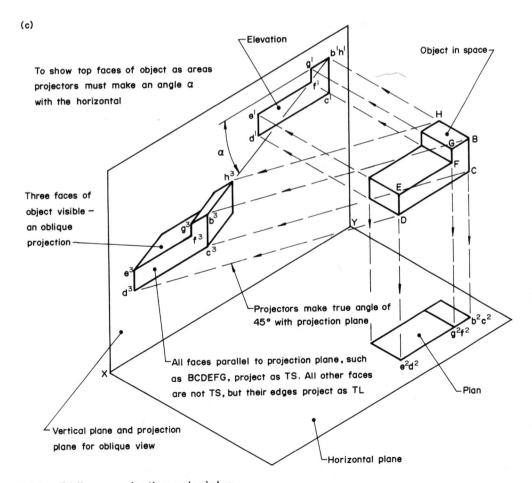

To show top faces of object as areas projectors must make an angle α with the horizontal

Elevation

Object in space

Three faces of object visible — an oblique projection

Projectors make true angle of 45° with projection plane

All faces parallel to projection plane, such as BCDEFG, project as TS. All other faces are not TS, but their edges project as TL

Plan

Vertical plane and projection plane for oblique view

Horizontal plane

Figure 7.11   *Oblique projection principles*

Self-assessment questions

7.9 What disadvantage of isometric drawings does oblique projection help to overcome?

7.10 Lines and faces on an object which are parallel to the plane of projection in an oblique drawing will appear as true lengths and true shapes respectively. True or false?

7.11 An oblique projection of an object may be drawn in which edges perpendicular to the plane of projection appear as true lengths. True or false?

Cavalier and cabinet projection

The oblique projection of the object in Figure 7.11(c) has been redrawn in Figure 7.12(a). Although this is a true projection, the object appears to be too thick. This is caused by no allowance being made for the apparent shortening and convergence of receding parallel lines. If the scale on the receding axis is reduced, the distortion becomes less noticeable; scales of three-quarters full size and half full size are commonly used. If full size is used on the receding axis the view is said to be in **cavalier projection**. When a scale of half full size is used the resulting view is a **cabinet projection**.

Variation of the angle of the receding axis

Different faces of the object may be shown in oblique projection by altering the angle which the receding axis makes with the horizontal. In Figure 7.13 the angles chosen emphasize the right-hand, left-hand, upper and lower faces of the objects. Easily drawn angles such as 30° and 45° should be used.

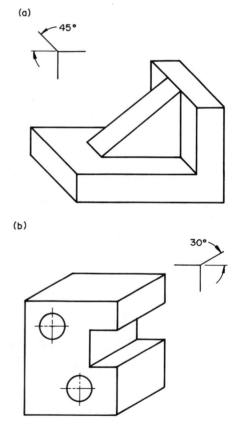

(a)

(b)

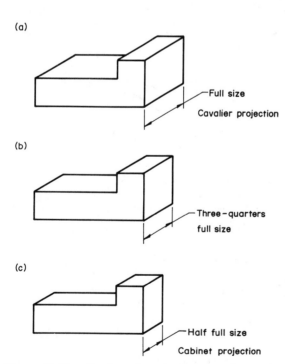

(a)

Full size
Cavalier projection

(b)

Three-quarters
full size

(c)

Half full size
Cabinet projection

Figure 7.12 *Reduced scale for receding axis to reduce distortion*

Figure 7.13 *Variation of angle of receding axis*

126

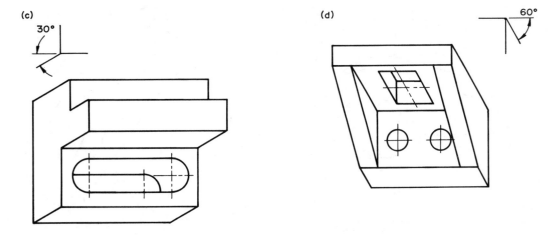

(c)

30°

(d)

60°

Varying the angle of the receding axis emphasises different faces of the object. 30°, 45° and 60° angles are commonly used

Figure 7.13 *Concluded*

Positioning the object relative to the projection plane

When the object is in some positions relative to the projection plane, the distorted appearance of the view may be objectionable. This is illustrated in the right-hand views in Figure 7.14. To minimise the distortion two rules should be observed. First, place the face of the object which contains circles or circular arcs parallel to the projection plane. Second, place the longest dimension of the object parallel to the projection plane. When these two rules conflict, as with long objects of circular cross-section similar

to that in Figure 7.14(a), the first takes precedence.

The first rule has been applied to the object shown in Figure 7.15. The faces containing the circular holes and the 60° angle have been positioned parallel to the projection plane so that they may be drawn as true size and true shape. The receding axis is drawn upwards to the right to show the 60° angle. Figure 7.15 illustrates also the use of a box which just encloses the object, as the first step in making the drawing. This technique is the same as that used in isometric projection.

Place face of object containing circles or arcs parallel to projection plane

Place longest dimension of object parallel to projection plane

(a)

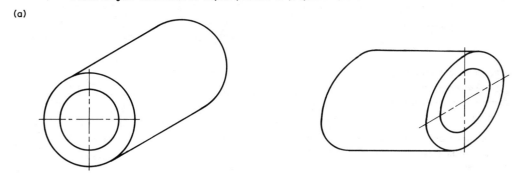

Figure 7.14 *Positioning object relative to the projection plane*

127

(b)

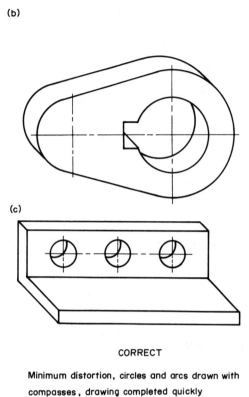

(c)

CORRECT

Minimum distortion, circles and arcs drawn with
compasses, drawing completed quickly

INCORRECT

Figure 7.14 *Concluded*

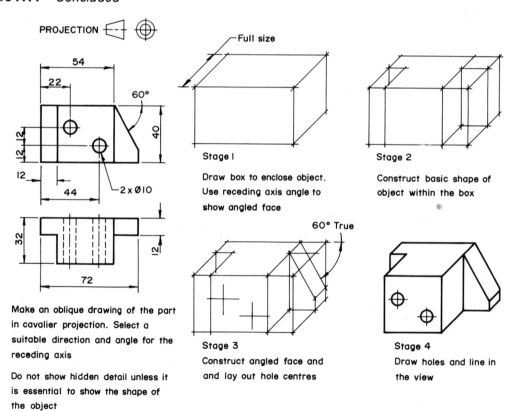

PROJECTION

54

22

60°

40

12 12

12

44

2 x Ø10

32

12

72

Make an oblique drawing of the part
in cavalier projection. Select a
suitable direction and angle for the
receding axis

Do not show hidden detail unless it
is essential to show the shape of
the object

Full size

Stage 1

Draw box to enclose object.
Use receding axis angle to
show angled face

Stage 2

Construct basic shape of
object within the box

60° True

Stage 3

Construct angled face and
and lay out hole centres

Stage 4

Draw holes and line in
the view

Figure 7.15 *Angles on a plane parallel to projection plane*

128

Self-assessment questions

7.12 What is the difference between cavalier and cabinet projections?

7.13 What two rules should be followed to minimise distortion in oblique views? Which rule takes precedence if they conflict?

7.14 If an object has circular holes in two faces at right angles to each other, an oblique projection of it cannot be drawn in which all the holes are true shape.   True or false?

Exercise 7.17   ## Exercise 7.18

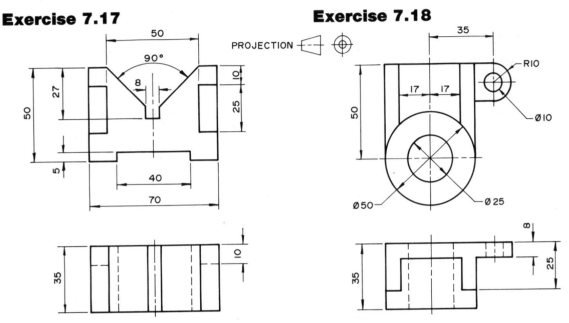

Make oblique drawings of the parts using cavalier projection.

Exercise 7.19   ## Exercise 7.20

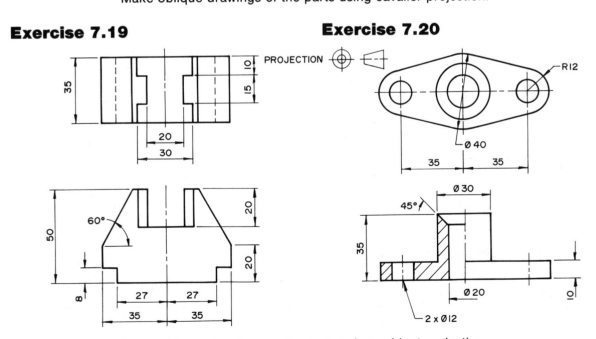

Make oblique drawings of the parts using cabinet projection.

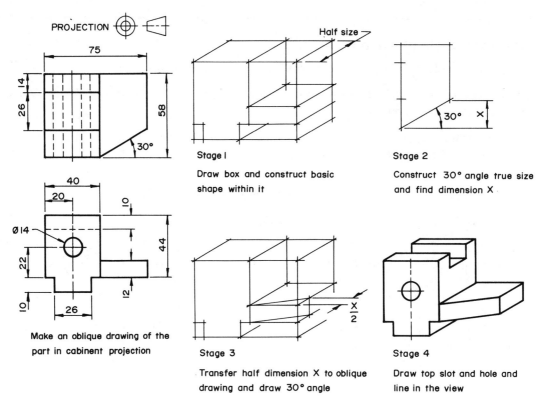

PROJECTION

75

14
26
58

30°

Make an oblique drawing of the part in cabinent projection

40
20
Ø14
22
10
44
12
10
26

Half size

Stage I

Draw box and construct basic shape within it

Stage 2

30° X

Construct 30° angle true size and find dimension X

Stage 3

X/2

Transfer half dimension X to oblique drawing and draw 30° angle

Stage 4

Draw top slot and hole and line in the view

Figure 7.16 *Angles on a receding plane*

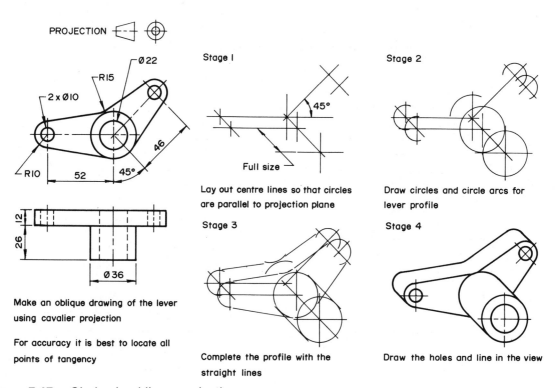

PROJECTION

Ø22
R15
2 x Ø10
46
R10
52
45°

12
26
Ø36

Make an oblique drawing of the lever using cavalier projection

For accuracy it is best to locate all points of tangency

Stage I

45°

Full size

Lay out centre lines so that circles are parallel to projection plane

Stage 2

Draw circles and circle arcs for lever profile

Stage 3

Complete the profile with the straight lines

Stage 4

Draw the holes and line in the view

Figure 7.17 *Circles in oblique projection*

Angles on a receding plane

In Figure 7.16 the 30° angle does not lie in a parallel plane to the 14 diameter hole, so it will not appear as its true size in the oblique drawing. However, from a true size view the dimension X can be transferred to the oblique drawing. Since cabinet projection is being used, X must be halved during transfer.

Circles in oblique projection

The object in Figure 7.17 is suited ideally for drawing in oblique projection. In isometric projection the circles and circular arcs would take much time to construct. In oblique projection they can be drawn directly with compasses, provided the object is positioned with them parallel to the projection plane.

With objects like this, instead of drawing a box it is more convenient to draw the centre lines first and use them as a skeleton on which to complete the view.

Exercise 7.21

Exercise 7.22

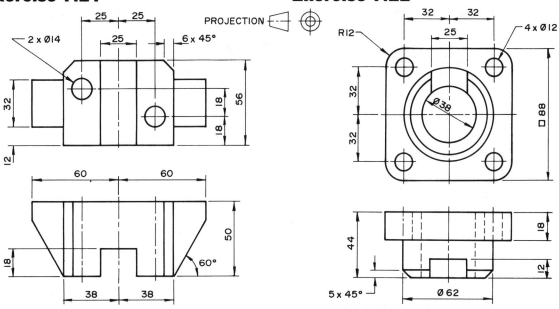

Make oblique drawings of the parts using cavalier projection.

Exercise 7.23

Exercise 7.24

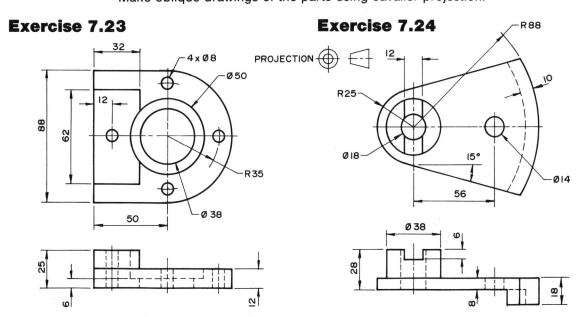

Make oblique drawings of the parts using cabinet projection.

Post test 7

1 (a) Drawings in orthographic projection are quicker to make than pictorial drawings.   True or false?

   (b) Pictorial drawings are more difficult to dimension than orthographic projection drawings. True or false?

   (c) The angle between any two axes on an isometric drawing is 135°.   True or false?

   (d) In an isometric drawing all the angles are true size.   True or false?

   (e) The main advantage of oblique projection is that curves parallel to the projection plane are true size.   True or false?

   (f) An oblique drawing with the receding axis full size is a cavalier projection.   True or false?

   (g) Hidden detail lines are usually omitted from pictorial drawings.   True or false?

2 Make isometric and cabinet projection drawings of the given part. Position the part to show it to the best advantage. Draw the isometric circle using the approximate method.

PROJECTION

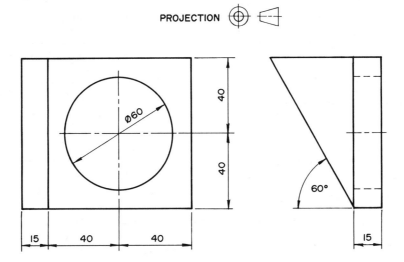

Electrical circuit diagrams

Principal learning objective

The student uses conventions shown in appropriate standards to interpret and draw electrical circuit diagrams.

Introduction

A drawing of a part made in orthographic projection enables a reader of the drawing to visualise the shape and proportions of the part completely. An electrical circuit diagram, on the other hand, does not show the shape of the parts of the circuit or their positions relative to each other. Instead, the parts of the circuit are represented by standardized graphical symbols. The way the symbols are linked to each other shows how the actual parts have to be joined with **conductors** for the circuit to function.

The conductors are usually of copper wire, insulated with plastic or rubber sleeving. They are fitted with lugs or **terminals** so that they can be attached to the parts of the circuit. Figure 8.1 shows three common types of terminal.

Figure 8.1(a) is an open-ended type, the sleeve of which is crimped to the end of the conductor after a short length of insulation has been stripped from it. That in Figure 8.1(b) has a washer end and the bared end of the conductor is soldered into the sleeve. Figure 8.1(c) shows a heavy terminal which is intended to be soldered to a thick conductor.

Conductors may also be metal strips on **circuit boards**. Components in the circuit are soldered to them using tags or wires which are part of the component.

A circuit diagram is drawn in sequence from left to right and/or from top to bottom. It should show the apparatus as it would be with no current flowing in it. Some items in the circuit are assemblies of parts which have a standard arrangement of the symbols representing the parts. Such arrangements should always be used when the assemblies appear in a more complex circuit diagram. Lines representing conductors should have as few changes of direction and cross-overs as possible. Earth and battery symbols can sometimes be repeated in order to simplify the arrangement of the diagram.

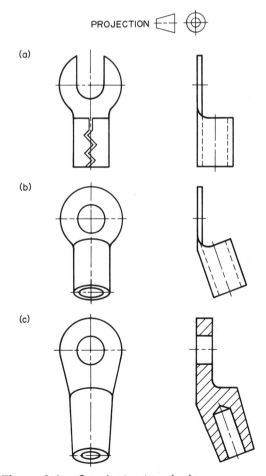

Figure 8.1 *Conductor terminals*

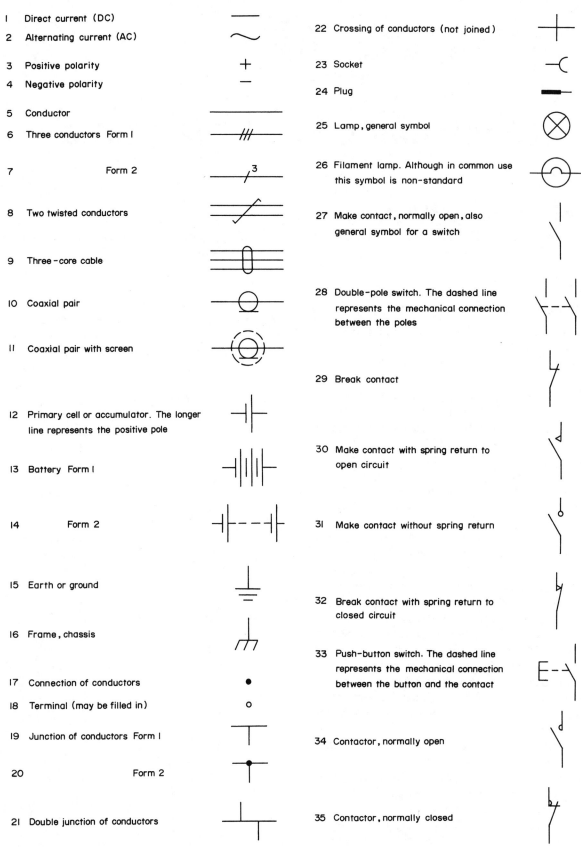

1   Direct current (DC)

2   Alternating current (AC)

3   Positive polarity

4   Negative polarity

5   Conductor

6   Three conductors Form I

7   Form 2

8   Two twisted conductors

9   Three-core cable

10 Coaxial pair

11 Coaxial pair with screen

12 Primary cell or accumulator. The longer line represents the positive pole

13 Battery Form I

14   Form 2

15 Earth or ground

16 Frame, chassis

17 Connection of conductors

18 Terminal (may be filled in)

19 Junction of conductors Form I

20   Form 2

21 Double junction of conductors

22 Crossing of conductors (not joined)

23 Socket

24 Plug

25 Lamp, general symbol

26 Filament lamp. Although in common use this symbol is non-standard

27 Make contact, normally open, also general symbol for a switch

28 Double-pole switch. The dashed line represents the mechanical connection between the poles

29 Break contact

30 Make contact with spring return to open circuit

31 Make contact without spring return

32 Break contact with spring return to closed circuit

33 Push-button switch. The dashed line represents the mechanical connection between the button and the contact

34 Contactor, normally open

35 Contactor, normally closed

Figure 8.2   *Selected electrical and electronic graphical symbols*

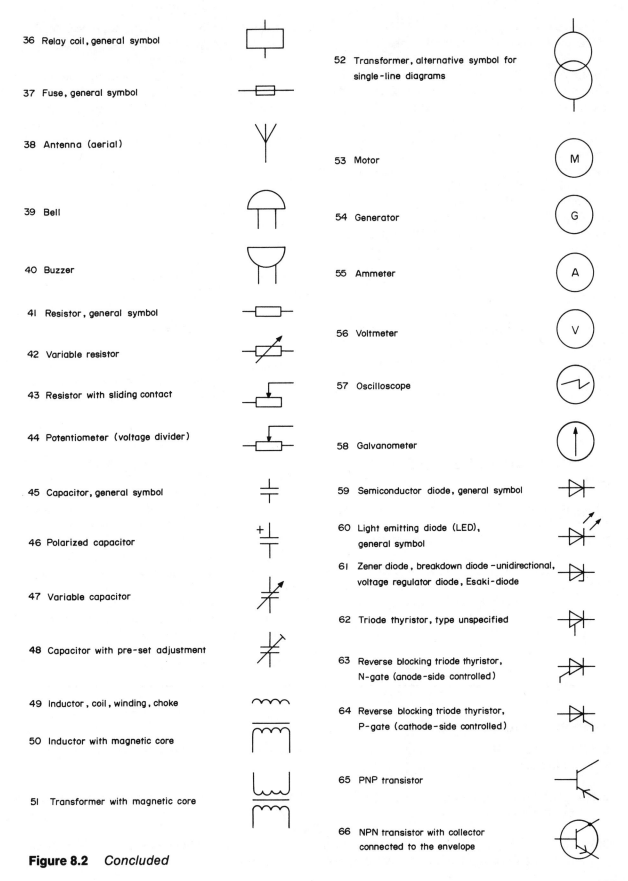

36 Relay coil, general symbol

37 Fuse, general symbol

38 Antenna (aerial)

39 Bell

40 Buzzer

41 Resistor, general symbol

42 Variable resistor

43 Resistor with sliding contact

44 Potentiometer (voltage divider)

45 Capacitor, general symbol

46 Polarized capacitor

47 Variable capacitor

48 Capacitor with pre-set adjustment

49 Inductor, coil, winding, choke

50 Inductor with magnetic core

51 Transformer with magnetic core

52 Transformer, alternative symbol for single-line diagrams

53 Motor

54 Generator

55 Ammeter

56 Voltmeter

57 Oscilloscope

58 Galvanometer

59 Semiconductor diode, general symbol

60 Light emitting diode (LED), general symbol

61 Zener diode, breakdown diode – unidirectional, voltage regulator diode, Esaki-diode

62 Triode thyristor, type unspecified

63 Reverse blocking triode thyristor, N-gate (anode-side controlled)

64 Reverse blocking triode thyristor, P-gate (cathode-side controlled)

65 PNP transistor

66 NPN transistor with collector connected to the envelope

Figure 8.2   *Concluded*

135

The student may find that different forms of some of the graphical symbols are used. Those shown in Figure 8.2 are the standard forms and have been selected from BS 3939: *Graphical symbols for electrical power, telecommunications and electronics diagrams*. The British Standards Institution has also published PP 7303: *Electrical and electronic graphical symbols for schools and colleges*, which is an abridgement of BS 3939. No attempt has been made here to show all the possible variants of all the graphical symbols in use today.

Symbols can be built up from standard elements. For example, the symbol for a variable resistor, number 42 in Figure 8.2, consists of the symbol for a resistor and the symbol for variability. If the resistor had a pre-set adjustment, then the symbol for this, illustrated in number 48 of Figure 8.2, would replace the variability symbol.

Another example is the symbols for diodes and thyristors, numbers 59 to 64 in Figure 8.2. These all consist of a basic symbol with modifications. In symbol 60, reversing the arrows would mean that the item was light dependent instead of light emitting. Note that these symbols must point in the direction of current flow.

Electrical quantities and their units

An electrical battery, or a generator, produces an **electromotive force** or e.m.f. which is measured in **volts**. The symbol for volts is V, and the unit symbol is V.

If the terminals of the battery are connected to the ends of a conductor, the e.m.f. will cause an electric current to flow in the conductor. The electric current is the movement of electrons through the conductor and the conventional direction of flow is from the positive pole of the battery to the negative pole. The current is measured in **amperes**, for which the symbol is I and the unit symbol A.

If the temperature is kept constant the current I in the conductor is directly proportional to the voltage V applied to it. This is **Ohm's law** and means, for example, that if the voltage is doubled, the current flowing is also doubled. Ohm's law can be expressed in symbols as:

$$I \propto V$$

which can be expressed mathematically as

$$\frac{V}{I} = \text{constant}$$

The constant is the **resistance** of the conductor, which is given the symbol R. Therefore

$$R = \frac{V}{I}$$

Resistance is measured in **ohms**, for which the unit symbol is Ω. This is the Greek letter 'omega'. Resistors are colour coded to show the value of their resistance.

If the positive terminal of a battery is connected to a plate P_1 and the negative terminal to a parallel plate P_2 with an insulator such as air between them, then a **capacitor** is formed. The e.m.f. of the battery causes the positive plate to have a higher electrical charge than the negative plate. If the insulator is by-passed by, for example, a switch connected across the plates, the capacitor will discharge.

When a capacitor is charged it is storing electricity and the greater its **capacitance** the more electricity it can store. The symbol for capacitance is C. It is measured in **farads** for which the unit symbol is F. However, a farad is a very large unit and for most practical purposes the microfarad, μF, is used.

Self-assessment questions

8.1   If a voltage of 9 V is applied to a conductor and a current of 2 A flows in it, what is the resistance of the conductor?

8.2   A conductor has a resistance of 20 Ω. What voltage must be applied to it to cause a current of 5 A to flow?

8.3   When a voltage of 12 V is applied to a conductor the current which flows is 3 A. What current will flow when the voltage is reduced to 10 V?

Series and parallel connection of resistors and cells

Series connection of resistors

Resistors connected in **series** are joined one after the other in the circuit, as shown in Figure 8.3(a). Therefore the current flowing is the same at all points in the circuit. It can also be shown that the total resistance of the circuit is the sum of the individual resistances. For Figure 8.3(a), if the total resistance is R, then

$$R = R^1 + R^2 + R^3$$

Parallel connection of resistors

In this method of connection, illustrated in Figure 8.3(b), corresponding ends of the resistors are joined together. Therefore the voltage is the same across all the resistors. The total current will divide so that some of it will flow through each resistor, how much depending on the resistance of the resistor. However, the total current will be equal to the sum of the currents in each resistor.

It can be shown that if the total resistance of the circuit in Figure 8.3(b) is R, then

$$\frac{1}{R} = \frac{1}{R^1} + \frac{1}{R^2} + \frac{1}{R^3}$$

Series connection of cells

When several cells are connected together, either in series or in parallel, a **battery** is formed. Figure 8.4(a) shows a battery made up of three cells in series. The positive pole of each cell is joined to the negative pole of the next, so producing a positive and negative terminal on the battery.

If the e.m.f. of the first cell is E^1 volts, of the second E^2 volts and the third E^3 volts, then the total voltage at the battery terminals will be $E^1 + E^2 + E^3$.

Parallel connection of cells

Figure 8.4(b) shows a battery made up of three cells connected in parallel. If the e.m.f. of each cell is E

(a)

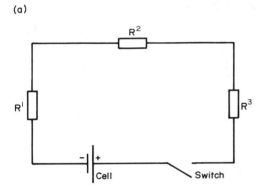

Resistors connected in series

(b)

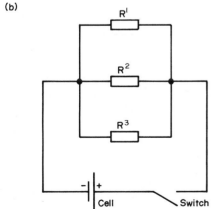

Resistors connected in parallel

Figure 8.3 *Series and parallel arrangements of resistors*

(a)

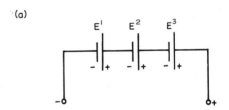

Battery with three cells in series

(b)

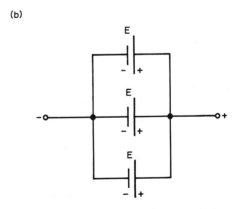

Battery with three cells in parallel

Figure 8.4 *Series and parallel arrangements of cells*

volts then the same current will flow in each of the cells. These currents will be only one-third of the current drawn from the battery. So a parallel connection of the cells would be used if the current required from the battery was more than any one cell could deliver safely. Also the total e.m.f. of the whole battery is then only the same as that of any one of the cells.

Self-assessment questions

8.4 A circuit consists of three resistors of 4 Ω, 8 Ω and 10 Ω connected in series. What is the total resistance of the circuit?

8.5 If the resistors in question 8.4 are connected in parallel, what will the resistance of the circuit be now?

8.6 Six 2 V cells are connected in parallel to form a battery. If the battery supplies 30 A to a circuit, what current will flow in each cell?

Exercise 8.1

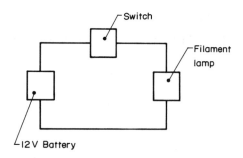

Draw the circuit using the correct graphical symbols in place of the boxes.

Exercise 8.2

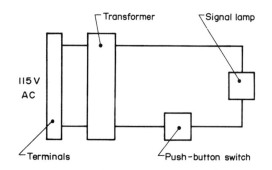

(a) Draw the circuit using the correct graphical symbols in place of the boxes.

(b) Redraw the circuit diagram twice:
   (i) with a buzzer in series with the lamp;
   (ii) with a bell in parallel with the lamp.

Exercise 8.3

A circuit consisting of three filament lamps, a switch and a fuse is to be supplied from a generator. Draw the circuit diagram so that if one lamp burns out the others will remain lit.

Exercise 8.4

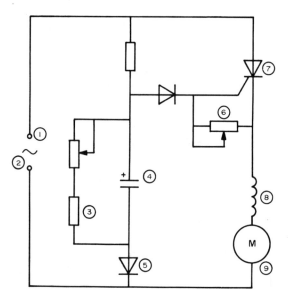

Identify the numbered items in the above circuit diagram.

Examples of simple electronic circuit diagrams

In Figure 8.5 the light emitting diode is designed to allow current to pass through it in the direction in which its symbol points. If the voltage were reversed, the current would flow in the opposite direction, which could cause damage to the diode. This can be prevented by connecting a second diode in parallel with the light emitting diode, arranged so that it will operate if the voltage is reversed.

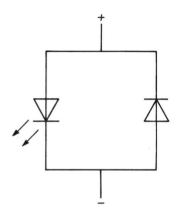

Figure 8.5 *Light emitting diode protected from reverse voltage*

The battery in Figure 8.6 is required to provide a constant voltage output, the magnitude of which is determined by the battery voltage less the voltage drop across the resistor. As the battery becomes discharged the output voltage will drop. This drop can be compensated for by adding a Zener diode, or constant-voltage diode, to the circuit as shown.

As in Figure 8.6, without the light-dependent resistor in Figure 8.7 the output voltage would be the difference between the battery voltage and the voltage drop across the fixed resistor. The light-dependent resistor acts like a variable resistor, the variability being caused by different light levels falling on it. When the light level is high the resistance is low. The output voltage is then also low because it is proportional to the resistance. When the light level decreases the resistance increases and so the output voltage increases.

The function of an electric circuit can often be shown more clearly if the **voltage rails** are drawn as short horizontal lines, as in Figure 8.8. Each line must be marked with its voltage.

In the figure, SK means socket; PL, plug; R, resistor; C, capacitor; and TR, transistor. The resistor values are in ohms and the capacitor values in microfarads.

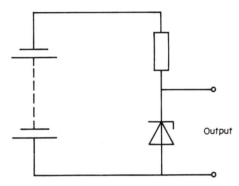

Figure 8.6 *Use of Zener diode to provide constant-voltage output*

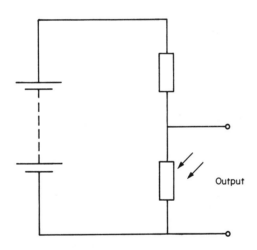

Output is small when light is high and greater when light is low

Figure 8.7 *Light-dependent resistor used as a potential divider*

Exercise 8.5

Draw the diagram for a test circuit to measure the current flowing through a resistor and the voltage across it. The supply to the circuit is through a transformer connected to 240 V AC mains and is controlled by a make contact switch.

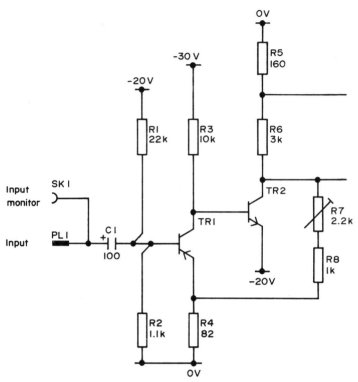

Figure 8.8 *Part of a video amplifier circuit diagram*

Exercise 8.6

A circuit supplied by a 12 V battery is required to operate a buzzer and a signal lamp independently through push-button switches. Draw the circuit diagram.

Exercise 8.7

Three bells in a circuit supplied by a 12 V battery are to be operated by a single make contact switch. Draw the circuit diagram.

Exercise 8.8

Two sets of four filament lamps are each operated by a make contact switch and protected by a fuse. If the supply is from 240 V AC mains, draw the circuit diagram.

Domestic electrical circuit diagrams

Figure 8.9 shows how an electricity supply is brought into a house. The electricity comes from a service cable, through a sealing box, to a fuse sealed by the electricity board. The live supply wire in the cable has red insulation and the neutral wire black. A third wire, the earth, is usually bare and connects the casing of the consumer's fuse box to ground.

From the service fuse the live and neutral wires pass through a meter which records how much electricity is used. From the meter they go to the mains switch, which is in the consumer's fuse box. From there power and lighting circuits are taken into the house.

Methods of wiring house circuits

Radial wiring. This method was used in houses

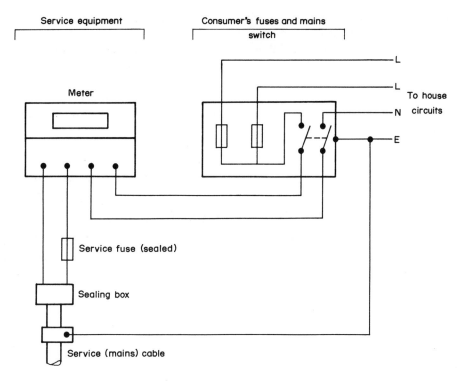

Figure 8.9 *Domestic electricity supply*

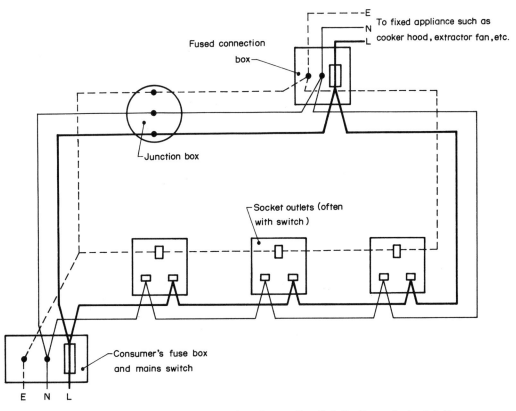

Figure 8.10 *Generalized ring-mains system for domestic distribution of electricity*

built before 1947. Electricity was supplied to each **socket outlet**, or power point, separately. Therefore, each socket outlet required its own fuse and these were all housed in one fuse box. This box had its own mains switch.

Separate circuits were installed for lighting and for equipment which needed large currents, such as cookers and water heaters. Each of these circuits had its own fuse box and mains switch.

Ring-mains wiring.   In houses built after 1947 the **ring-mains** system has been used. A generalised diagram for a ring-mains power circuit is shown in Figure 8.10. The socket outlets, and any **fused connection boxes** that are needed, are connected by a continuous loop of cable, called a ring circuit. Usually there is one ring to supply the socket outlets on each floor. Lighting and heavily loaded appliances each have their own ring and their own fuses in the

mains fuse box, but generally there is only one mains switch.

Branch cables, called **spurs**, are connected to the ring cable to supply any socket outlets that cannot conveniently be reached by the ring. However, there are limits to the number of spurs which can be installed, and to the number of socket outlets on each spur.

Graphical symbols for architectural installation plans and diagrams

In Figure 8.11 is shown a selection of graphical symbols which are used by architects to represent the electrical services in buildings. Symbols from Figure 8.2, such as those for conductors and cables, are used where appropriate.

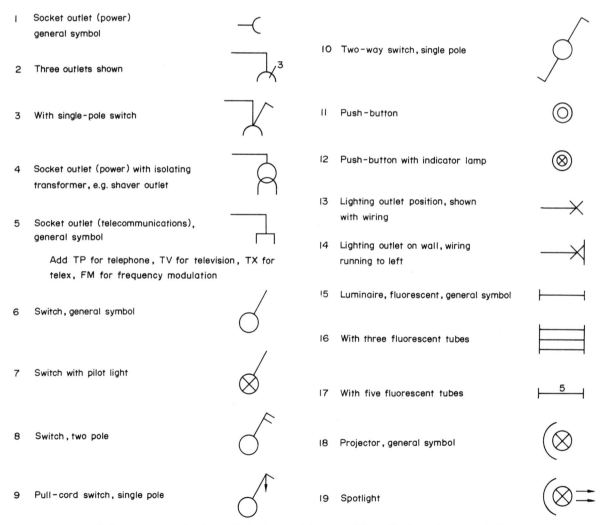

Figure 8.11   *Selected graphical symbols for architectural installation plans and diagrams*

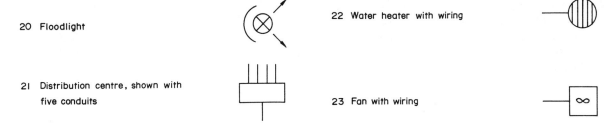

20 Floodlight

21 Distribution centre, shown with
   five conduits

22 Water heater with wiring

23 Fan with wiring

Figure 8.11 *Concluded*

Exercise 8.9

The circuit for a television aerial has the following parts: the aerial, to which a length of screened coaxial cable is connected by a plug and socket; at the other end of the coaxial cable is a socket outlet for connecting the television set. Using standard graphical symbols draw the circuit diagram.

Exercise 8.10

Part of the lighting circuit in a house consists of a pendant light controlled by a wall switch. Using the correct graphical symbols draw the circuit diagram, beginning at the mains switch.

Exercise 8.11

Two individually switched spot lights are to be installed on the wall of a room. Beginning at the mains switch, draw the circuit diagram with the correct graphical symbols.

Post test 8

1 Identify the numbered items in the car horn circuit diagram.

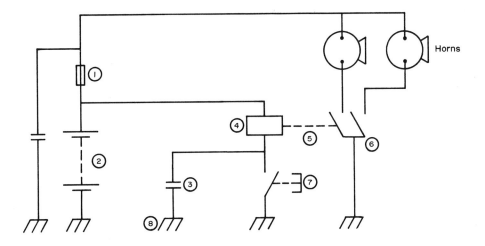

2 In a warning circuit a SIX CELL BATTERY has connected in series to the positive pole a FUSE followed by a MAKE CONTACT SWITCH and a test AMMETER. To the negative pole of the battery are connected two RESISTORS in series, the second of which is variable. A test VOLTMETER is connected in parallel across the circuit. A CHANGE OVER SWITCH energises either a BELL or a SIGNAL LAMP. Draw the circuit diagram using standard graphical symbols.

3

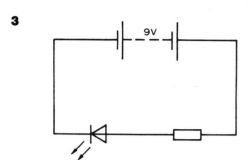

The allowable voltage across the LED is 2 V and the current in the circuit is 10 mA. Work out the resistor value.

4 Draw the following graphical symbols for use on architectural drawings:
(a) Three-core cable   (b) Double power socket outlet with single pole switch   (c) Telephone outlet   (d) Two pole switch   (e) Single pole, two way switch   (f) Two pole, pull-cord switch with pilot light   (g) Lighting outlet on wall   (h) Fluorescent luminaire with four tubes   (j) Floodlight (k) Fan   (l) TV aerial   (m) Water heater

Answers to self-assessment questions

Chapter 1

1.1 Orthographic projection is easier to draw than pictorial projections and gives accurate information on the size and shape of an object without using words.

1.2 An orthogonal projector is one that is perpendicular to the plane on which the view of the object is projected.

1.3 A square must be parallel to a plane for its projection on the plane to be a square.

1.4 a) True.
b) False. The views are the same but their positions relative to each other are different.
c) True.
d) True.

1.5 Visible edges are represented by continuous thick lines and hidden edges by thin dashed lines.

1.6 True.

1.7 False. A continuous thin line is used where the circle representing the hole is below about 12 diameter.

1.8 False. The minimum of hidden detail should be shown, consistent with describing the object completely.

1.9 False. No orthogonal projection of a line can be longer than the true length of the line.

1.10 The plan.

1.11 The elevation.

1.12 True.

1.13 False. The projectors for the auxiliary view must be at right angles to an edge view of the surface.

1.14 True.

Chapter 2

2.1 A cutting plane is represented by a thin chain line, thickened at the ends and at any changes of direction.

2.2 False. All visible lines behind the cutting plane must be shown on a full sectional view.

2.3 False. Hatching may be omitted when the meaning of the drawing is clear without it.

2.4 True.

2.5 No. With half sections the cutting plane passes half-way through the object, and with a local section it passes part-way through the object.

2.6 Half sections save time and space in making a drawing.

2.7 The boundary of a local section is a thin line.

2.8 The outline of a revolved section is a thin line and that for a removed section is a thick line.

2.9 No, any which are unessential for understanding the drawing may be omitted.

2.10 Yes, or it may be difficult to decide how the cutting plane passes through the object.

2.11 False. If the sectional view is projected onto the horizontal or vertical planes it is a principal view.

2.12 False. The boundary is an outline.

2.13 A sectioned shaft is completely hatched when the section plane cuts it transversely, giving a circular sectional view.

2.14 The advantages of using an aligned section are that awkward projections are avoided and features on the cutting plane are shown at their true distances from the centre of the part.

Chapter 3

3.1 False. Projection and dimension lines are thin.

3.2 The usual angle between projection and dimension lines is 90°.

3.3 The recommended maximum and minimum lengths for arrow heads are 5 mm and 3 mm.

3.4 Leader lines ending in a dot should finish inside the outline of the view.

3.5 False. Horizontal dimensions are arranged to be read from the bottom of the drawing.

3.6 An auxiliary dimension is indicated by being enclosed in parentheses.

3.7 When chain dimensioning is used, tolerances on the dimensions in the chain will accumulate which may affect the correct functioning of the part.

3.8 $6 \times \varnothing 16 \times 25$ LG SQ HD

3.9 AF = Across flats;   HEX HD = Hexagon head;   INSUL = Insulation or insulated;   RH = Right hand;   HYD = Hydraulic;   LG = Long;   SH = Sheet;   THD = Thread;   SR = Spherical radius;   CRS = Centres.

3.10 False. Circles should be dimensioned using their diameters.

3.11 (a)

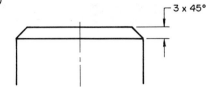

(b)

3.12

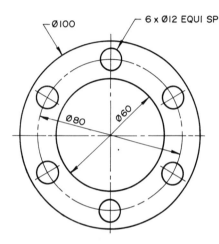

Note. Other arrangements of the dimensions are possible

3.13

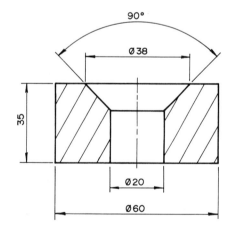

3.14 False. Auxiliary dimensions are untoleranced.

3.15 20.1

3.16 25 MIN has a unilateral tolerance, 35.5 ± 0.5 has a bilateral tolerance.

3.17
(a) $61° \begin{smallmatrix} 0 \\ -1° \end{smallmatrix}$ or $60° \begin{smallmatrix} +1° \\ 0 \end{smallmatrix}$;
(b) 60° 30′ ± 0° 30′ or 60.5° ± 0.5°

3.18 False.

3.19 No. If the dimension is altered one may be missed which would give two different sizes for the dimension.

3.20 Yes. It avoids the reader having to consult two or more views to find all the dimensions.

3.21 Size dimensions: 10 diameter; 15 deep; 60 diameter. Location dimension: 80. Note that 'equally spaced' is also a location instruction.

Chapter 4

4.1 True.

4.2 False. Generally the tapping drill depth is greater than the useful length of thread.

4.3 False. The included angle of a tapping drill point is always shown as 120°.

4.4 True.

4.5 M = ISO metric thread; 20 = major diameter; 1.5 = pitch; 5H = tolerance class symbol. The thread is internal because the H in the tolerance class symbol is a capital letter. The thread is in the fine pitch series because the pitch is stated.

4.6 1.5 times the major diameter of the thread.

4.7 0.8 times the major diameter of the thread.

4.8 A screw is threaded for virtually its entire length but a bolt is threaded for only part of its length.

4.9 False. A plain portion in the middle of the stud separates the nut end and the metal end.

4.10 Studs and nuts are used where there is not enough space on one side of the joint for bolt heads or nuts, and where parts have to be removed and replaced frequently.

4.11 A plain washer is used under a nut to prevent damage to the surface of the part as the nut is tightened.

4.12 Screws with hexagon socket heads are tightened with an Allen key.

4.13 Frictional locking devices: lock nuts; locking washers; stiff nuts; grub screws. Positive locking devices: slotted nuts; castle nuts; taper pins; locking plates; tab washers; wire locking.

4.14 A star washer is a frictional locking device.

4.15 A castle nut has a collar at the top through which the split pin passes.

4.16 A castle nut is safer than a slotted nut. The legs of the split pin are bent round the collar and so do not stand out as they do with slotted nuts.

4.17 The vee in the locking tab on tab washers enables the tab to be bent up against two faces of the nut.

4.18 The small leg on some tab washers fits in a hole in the part, preventing the washer rotating as the nut is tightened.

4.19 A siamese tab washer is illustrated in Figure 4.8(h).

4.20 The wire should run through the bolt heads in such a way that any tendency for the bolts to loosen tightens the wire.

Chapter 5

5.1 Rivet heads used in heavy-gauge work are illustrated in Figure 5.1.

5.2

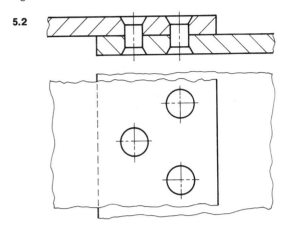

Double riveted lap joint with diagonal riveting

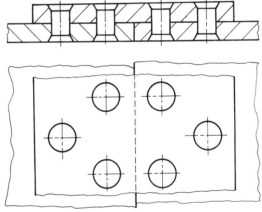

Double riveted butt joint, diagonal riveting, single cover plate

5.3 A light-gauge joint with a snap head rivet is shown in Figure 5.4(b).

5.4 Tubular or hollow rivets can be used in light-gauge work where only one side of the joint is accessible.

5.5 A joint made with a Tucker 'Pop' rivet is illustrated in Figure 5.6(a).

5.6 A key is fitted between a hub and a shaft to prevent the parts rotating relative to each other.

5.7 Figure 5.7(b) shows a rectangular taper key with a gib head.

5.8 Advantage: self-adjusts to any taper in the hub keyway. Disadvantages: cannot be used as a feather; shaft weakened by deep keyway.

5.9 Screw fasteners are designed generally to take tensile loads. Dowels are designed to take shear loads.

5.10 Holes for dowels are drilled together to ensure the parts are located accurately relative to each other.

Chapter 6

6.1 Conventional practices and representations save time in making drawings and save space on the drawings.

6.2 Quarter views could be used for objects which are symmetrical about two centre lines at right angles.

6.3 The symmetry convention can be used on asymmetrical views provided the asymmetric half is drawn and the asymmetrical features are identified.

6.4 An interrupted view of a solid shaft is shown in Figure 6.4(a).

6.5

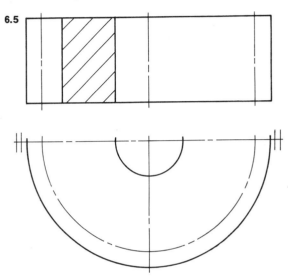

6.6 Both conventional representations use thin, crossed, diagonal lines.

6.7 Figure 6.10(a) shows the conventional representation of the outside view and sectional view of a cylindrical helical compression spring.

6.8 The machining symbol which is used when a surface is not to be machined is shown in Figure 6.11(c).

6.9 The machining symbol required is similar to that in Figure 6.11(e).

6.10 The machining symbol required is illustrated in Figure 6.11(f), except that the maximum surface value is to be added.

6.11 The required welding symbols are shown in Figure 6.13(b), (j), (l) and (e).

6.12 (a) (b) (c) (d)

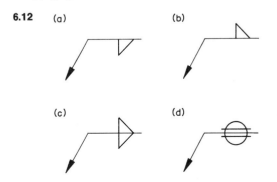

Chapter 7

7.1   Advantage: can be made and read without a knowledge of orthographic projection. Disadvantages: some pictorial projections are badly distorted; take longer to make than the same drawing in orthographic projection; are difficult to dimension; in some pictorial projections, circles and curves take a long time to draw; in pictorial projections some lines cannot be measured.

7.2   120°.

7.3   The drawing of an isometric view is helped by first drawing a box into which the object will just fit.

7.4   True.

7.5   Hidden detail is usually omitted from isometric drawings because its use would make the drawing difficult to read.

7.6   True.

7.7   False. The circular arcs in the four-centre method only approximate to a true ellipse.

7.8   True.

7.9   In oblique projection some circles on the object appear as circles. In isometric projection all circles on the object appear as ellipses.

7.10   True.

7.11   True, if the scale on the receding axis is full size.

7.12   In cavalier projection the scale on all three axes is full size. In cabinet projection the scale on the receding axis is half full size.

7.13   a) The face of the object which contains circles or circular arcs should be parallel to the projection plane.

b) The longest dimension of the object should be parallel to the projection plane.
Rule (b) takes precedence if the rules conflict.

Chapter 8

8.1 $4.5 \, \Omega$

8.2   100 V

8.3   2.5 A

8.4 $22 \, \Omega$

8.5 $2.11 \, \Omega$

8.6   5 A

Solutions to selected exercises

Exercise 1.1

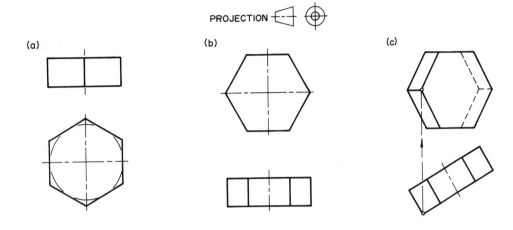

Exercise 1.4

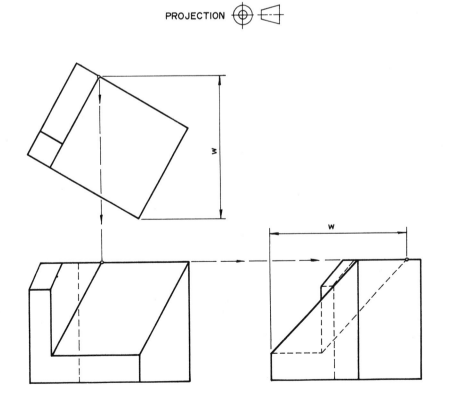

Exercise 1.8

PROJECTION

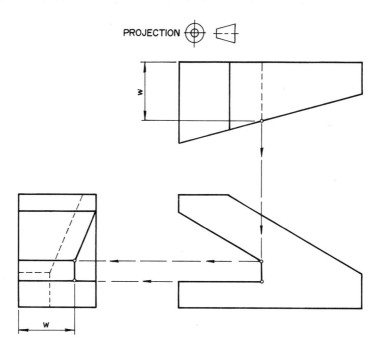

Exercise 1.10

PROJECTION

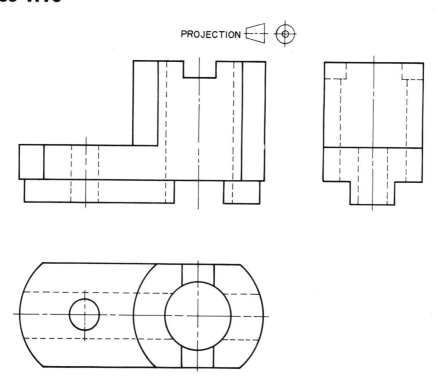

Exercise 1.13

PROJECTION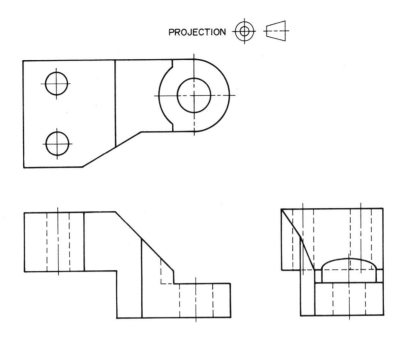

Exercise 1.17

PROJECTION

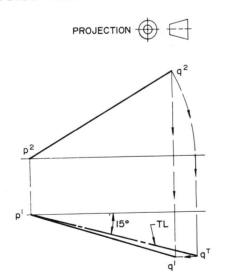

Exercise 1.21

PROJECTION

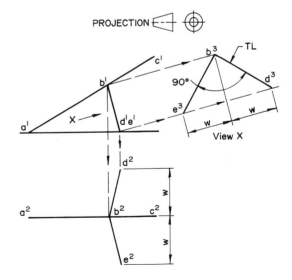

Exercise 1.24

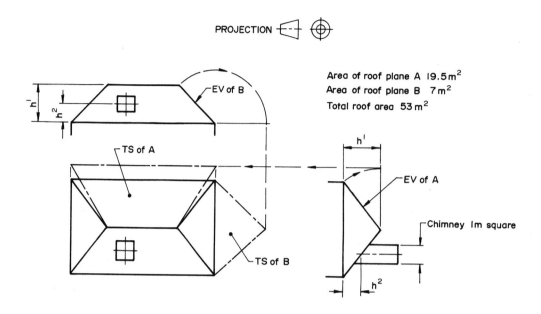

PROJECTION

Area of roof plane A 19.5 m^2
Area of roof plane B 7 m^2
Total roof area 53 m^2

EV of B

TS of A

TS of B

EV of A

Chimney 1m square

Exercise 1.26

PROJECTION

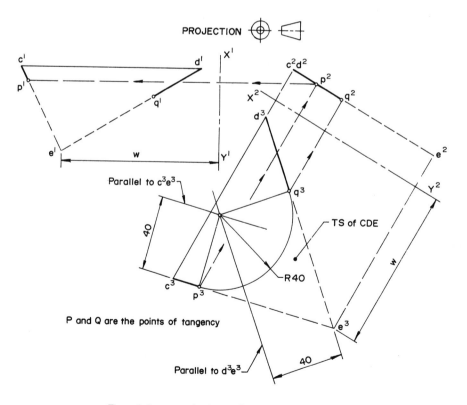

Parallel to c^{3}e^3

TS of CDE

R40

P and Q are the points of tangency

Parallel to d^{3}e^3

The solution may also be obtained using the revolution method

152

Exercise 1.28

PROJECTION ⊕ ⊏⊐

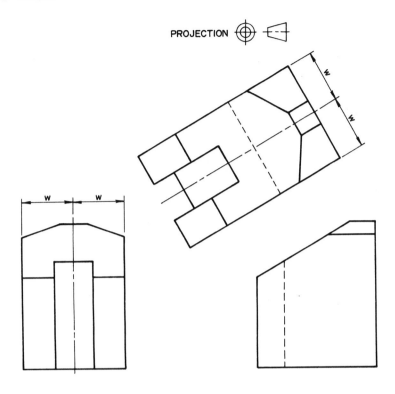

Exercise 1.31

PROJECTION ⊏⊐ ⊕

Face A

TS of face A

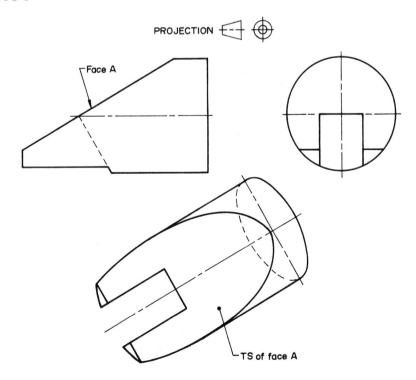

Exercise 2.2 solution

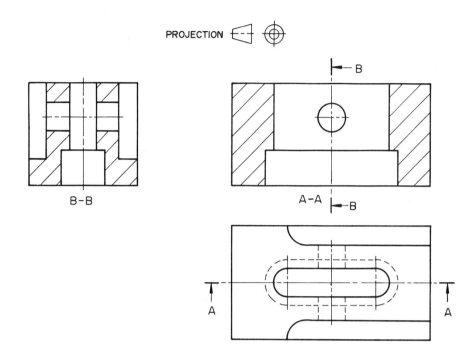

B-B

A-A

Exercise 2.5 solution

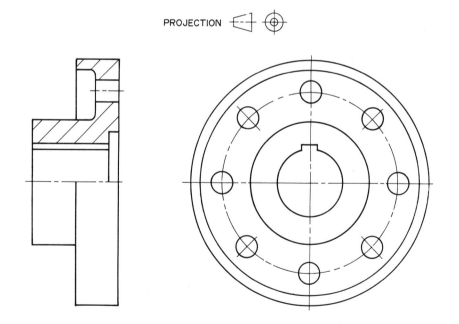

Exercise 2.10

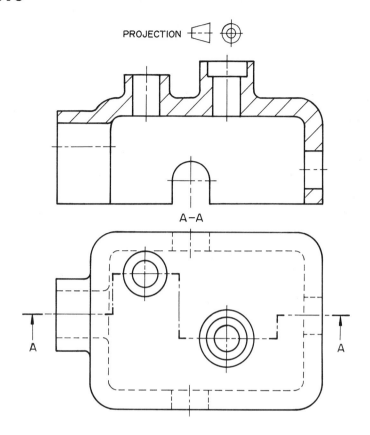

PROJECTION

A–A

A   A

Exercise 2.13

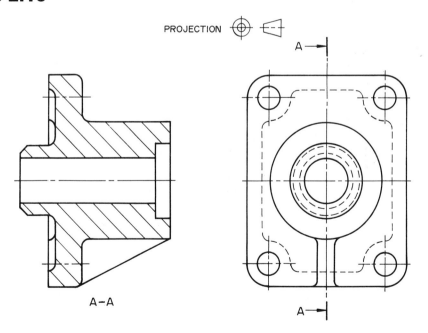

PROJECTION

A

A–A

A

Exercise 3.3

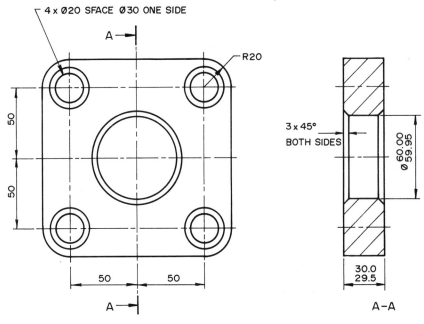

4 x Ø20 SFACE Ø30 ONE SIDE

A

R20

50
50

50   50

A

3 x 45°
BOTH SIDES

Ø 60.00
59.95

30.0
29.5

A-A

TOLERANCES EXCEPT WHERE STATED: LINEAR ± 0.2 , ANGULAR ± 0° 30'

Exercise 3.6

(a) The datum face is A

(b) Face A to face D is 37.67 / 38.33
Face B to face C is 21.67 / 22.33
Face C to face E is 51.84 / 52.16

(c) Maximum bore Ø 40.055
Minimum bore Ø 39.996

Exercise 3.7

(a) Thickness of spacer 4.70 / 4.77

(b) Length of gear hub 60.08 / 60.63

Exercise 3.8

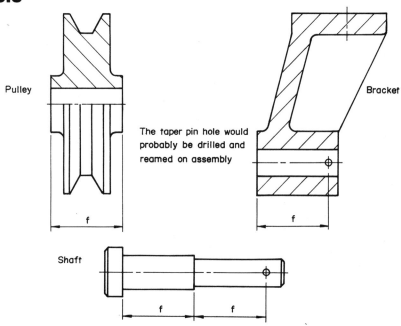

Pulley

The taper pin hole would
probably be drilled and
reamed on assembly

Bracket

f

f

Shaft

f   f

Exercise 4.2

PROJECTION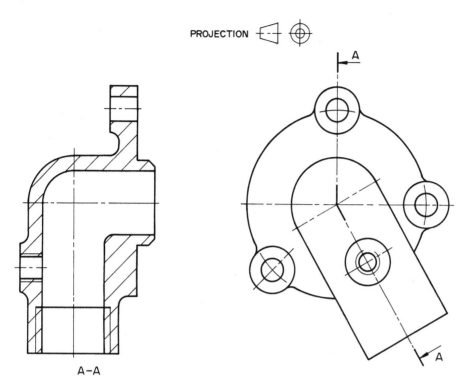

A–A

Exercise 4.4

PROJECTION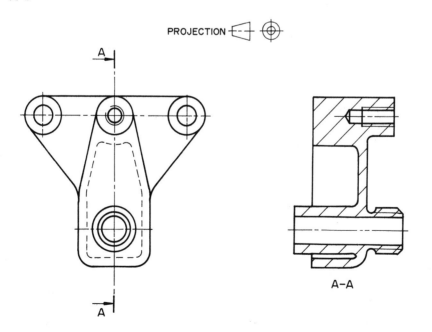

A–A

Exercise 4.8

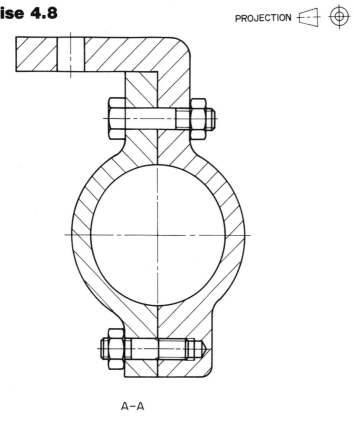

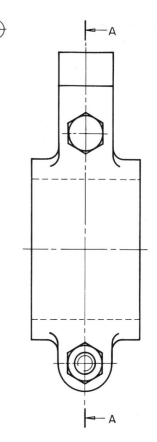

A–A

Exercise 4.10

PROJECTION ⊏⊐ ⊕

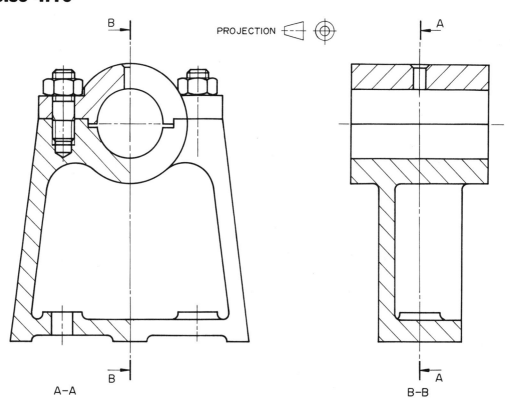

A–A

B–B

Exercise 6.2

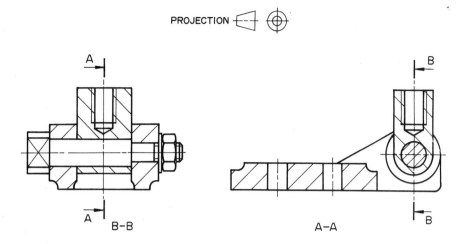

PROJECTION

A-A   B-B

Exercise 6.3

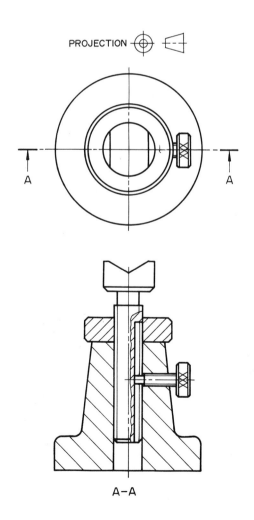

PROJECTION

A-A

159

Exercise 6.5

A

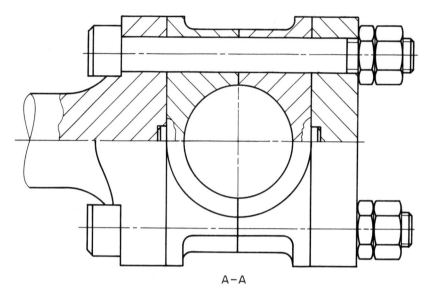

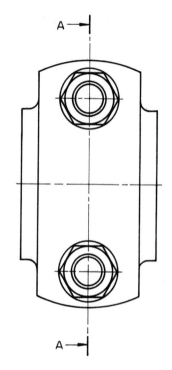

A–A

A

Exercise 6.8

PROJECTION

A

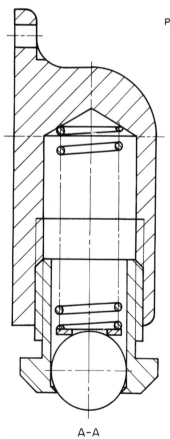

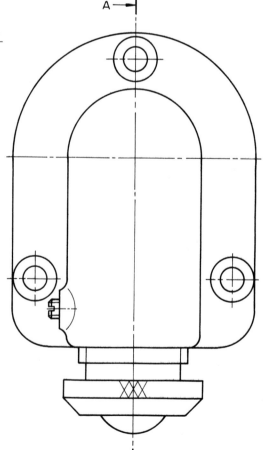

A–A

A

160

Exercise 7.2

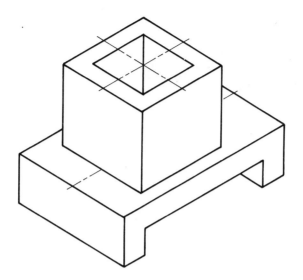

Exercise 7.5

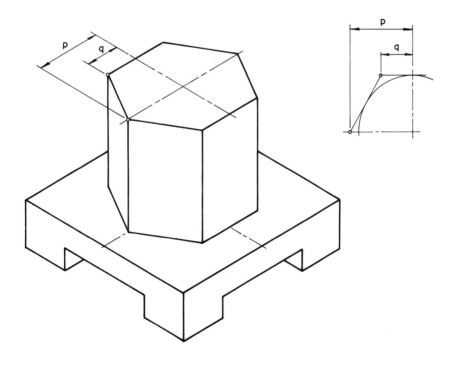

Exercise 7.8

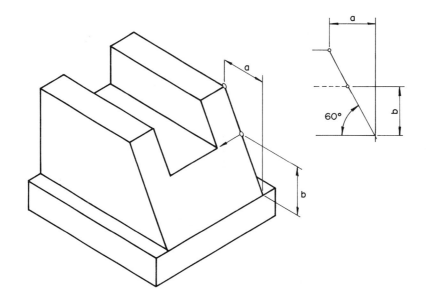

Exercise 7.12

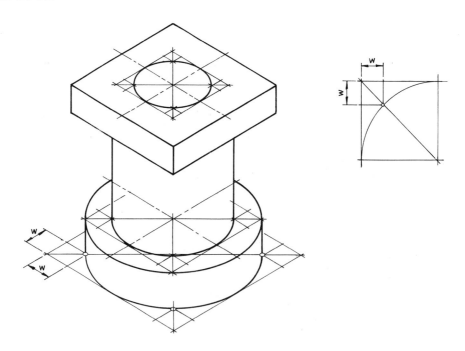

Exercise 7.15

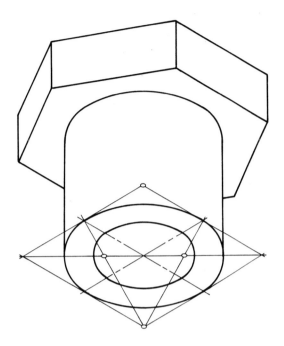

Exercise 7.18

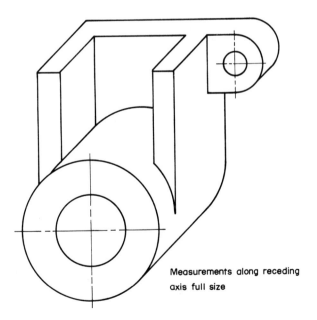

Measurements along receding
axis full size

Exercise 7.22

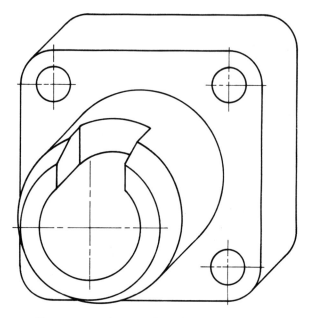

Measurements along receding axis full size

Exercise 7.24

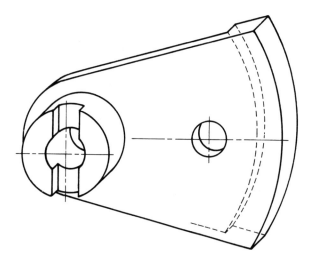

Measurements along receding axis half full size

Hidden detail clarifies shape of object

Exercise 8.2

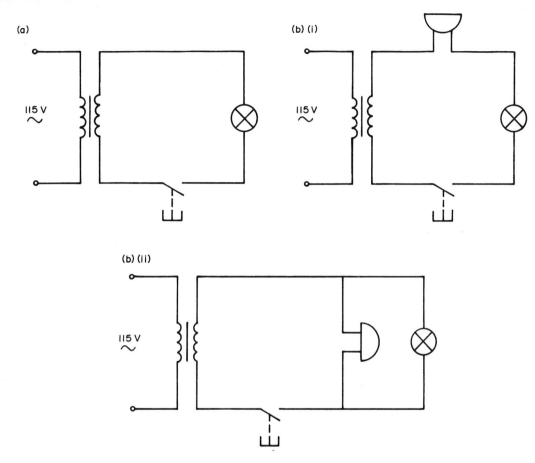

(a)

115 V
~

(b) (i)

115 V
~

(b) (ii)

115 V
~

Exercise 8.7

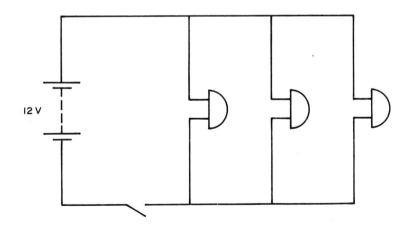

12 V